Entropy in relation to incomplete knowledge

# Entropy in relation
# to incomplete knowledge

**K. G. DENBIGH**

*Honorary Research Fellow, Chelsea College*
*University of London*

**J. S. DENBIGH**

*Mathematician, St George's Hospital Medical School, London*

The right of the
University of Cambridge
to print and sell
all manner of books
was granted by
Henry VIII in 1534.
The University has printed
and published continuously
since 1584.

CAMBRIDGE UNIVERSITY PRESS

*Cambridge*

*London New York New Rochelle*

*Melbourne Sydney*

*7291- 0999*

**PHYSICS**

Published by the Press Syndicate of the University of Cambridge
The Pitt Building, Trumpington Street, Cambridge CB2 1RP
32 East 57th Street, New York, NY 10022, USA
10 Stamford Road, Oakleigh, Melbourne 3166, Australia

© Cambridge University Press 1985

First published 1985

Printed in Great Britain at the University Press, Cambridge


Library of Congress catalogue card number: 84-23108

*British Library Cataloguing in Publication Data*
Denbigh, K.G.
Entropy in relation to incomplete knowledge.
1. Entropy
I. Title   II. Denbigh, J.S.
536'.73   QO318.E57

ISBN 0 521 25677 1

MP

# CONTENTS

# PREFACE

Since the early days of statistical thermodynamics the idea has become widely accepted that entropy really signifies nothing more than a lack of human knowledge. This view has been strengthened by the advent of information theory where entropy is specifically equated with an inverse measure of 'information'. This, in its turn, has led to statements that entropy is subjective, that it is an anthropomorphic concept.

Although these ideas are by no means recent, the time is appropriate to examine them much more critically than has been attempted hitherto. For it would appear that the same, or similar, arguments might affect other branches of science as well as thermodynamics and statistical mechanics. Indeed, there are well-known aspects of quantum mechanics which have suggested to some theorists that 'the observer' is perhaps not fully separable from what he observes.

Is it the case perhaps that science can no longer be regarded as providing a fully objective account of natural phenomena? This question raises some very profound epistemological and ontological issues concerning the scientific enterprise in general. The aim of this book is to consider these issues in that area in which they first arose, namely in thermodynamics and statistical mechanics. Quantum theory will be referred to only in so far as it has a bearing on this restricted field of inquiry.

We are much indebted to Dr Michael Redhead for reading an early version of the section on quantum mechanics. We have also greatly benefited from discussions with Professor Post, Dr Redhead, Dr Lavis, Mr Steven French and others at Chelsea College.

*London, April 1984*

# 1

## *Is entropy subjective in thermodynamics?*

### 1.1. Introduction

Nearly two centuries elapsed between the firm establishment of mechanics as a branch of science and a comparable stage in the development of thermodynamics! During that period, as we know, there were some studies of particular instances of irreversible phenomena such as heat conduction and the diffusion of gases. Newton himself remarked (1730, p. 373), in a telling phrase, that 'Motion is more apt to be lost than got, and is always upon the Decay.' But there was little attempt to view the various irreversible processes in relation to each other, or to develop a comprehensive theory embracing both the conservation of energy and its universal 'dissipation'. This had to await the work of Carnot, Helmholtz and Joule, and the subsequent systematisation due to Clausius and Kelvin in the mid 19th century.

In this book we shall be concerned with some of the consequences of the attempts to understand, or to explain, the newly arrived science of thermodynamics on the basis of the older science, mechanics. These attempts brought up some deep epistemological and metaphysical issues and, in particular, gave rise to a subtle transference of ideas from 'randomness' to 'lack of knowledge', and from there to 'subjectivity'. To put this point more specifically, there developed a marked tendency among the 19th-century scientists to attribute any apparent randomness in natural phenomena to a lack of sufficient knowledge about those phenomena rather than to any real chance element in nature. And there remains at the present time a strongly entrenched view to the effect that entropy is a subjective concept precisely because it is taken as a measure of 'missing information' – information which we *might* use but don't, due to thermodynamic systems being incompletely specified.

Before dealing with the merits of these issues, it may be as well to substantiate the foregoing with a brief historical account and quotations from some of the leading scientists, such as Maxwell and Gibbs.

It will be recalled that in 1867 Maxwell corresponded with Tait about the 'demon' who could supposedly defeat the Second Law by having a sufficiently detailed perception of the movements and velocities of individual molecules. This led Maxwell to the modern view that the Law is not absolute but is statistical. For instance, in his *Theory of Heat* (1875, p. 153), he remarked that the status of the Law would be very different if we had the powers of 'a being whose faculties are so sharpened that we can follow every molecule in its course'. Failing such powers, he went on to say, we are compelled to adopt 'the statistical method of calculation'.

This quotation has an important bearing on the two-stage 'transference of ideas' which was mentioned above. Let us consider these stages in turn.

Maxwell's use of the word 'statistical', pointed out by Brush in his historical study (1976), did not imply that Maxwell believed in the objective occurrence of randomness in the molecular motions: 'For Maxwell it is our knowledge of the world that is statistical, not the world itself . . . .' Although Maxwell, and Boltzmann too, used such terms as 'randomness' and 'disordered motions', Brush went on to say that it was nevertheless difficult for them, as for other scientists of the period, 'to abandon the view of Laplace that one assumes phenomena to be random merely because of lack of knowledge rather than because of any inherent indeterminism'. In a similar comment, Thomas Kuhn (1978, p. 45) remarked that Boltzmann, in his 1872 paper on the *H*-theorem, regarded the molecular distribution function as fully *determining* the later state of the gas. It was only as the result of Loschmidt's challenge, said Kuhn, that Boltzmann gradually became aware of the real significance of statistical treatments. Even so, the possibility that 'randomness' and 'chance' may be fully objective features of natural phenomena, and not simply a matter of lack of knowledge, was perhaps not widely accepted until after the later forms of the quantum theory had been developed in the 1920s. This point is also well made by Rosenfeld (1953).

The second stage of transference – from 'lack of knowledge' to 'subjectivity' – is also to be seen in the foregoing passage from the *Theory of Heat*. Although Maxwell was entirely right in his view that the Second Law is statistical (as was later shown by the existence of fluctuation phenomena, such as the Brownian motion), the subjective features of the argument by which he arrived at this conclusion were seriously misplaced. Maxwell emphasised the imperfection of *human* powers in comparison with those alleged to be available to the demon. But with the coming of quantum theory this was seen to have been a false step – for it neglected the minimum energy transfer of one quantum, of frequency sufficiently greater than that of the black body radiation within the enclosed gas, which is necessary to any physical entity, whether man or demon, if it is to be capable of detecting the moving molecule. And it is precisely this energy transfer which 'saves' the Second Law – although only 'on the average' in the statistical sense.

Although Maxwell could not have known about this factor his defective argument may well have been highly influential, in view of his great prestige, in paving the way to the subjective notion of entropy.

A further example of this kind occurs in Maxwell's article, 'Diffusion', in the ninth edition of the *Britannica*. At that time Gibbs had already described what came to be known as his 'Paradox' concerning the entropy of mixing of perfect gases. However similar or dissimilar these gases are, the entropy of mixing is exactly the same, but if the gases are *identical* the entropy of mixing changes discontinuously to zero. Maxwell instanced the case where two samples of gas previously supposed to be identical were subsequently found to be different. As he said, we must now attribute a dissipation of energy to their mixing where previously we attributed none. He continued:

It follows from this that the idea of dissipation of energy depends on the extent of our knowledge .... Dissipated energy is energy which we cannot lay hold of and direct at pleasure, such as the energy of the confused agitation which we call heat. Now confusion, like the correlative term order, is not a property of things in themselves, but only in relation to the mind which perceives them.

Just how the peculiar discontinuity which is involved in Gibbs' Paradox can be resolved is a separate issue which Maxwell did not attempt to deal with. What is surprising in Maxwell's remarks is a certain *non sequitur*. Although we can accept that 'confusion' and 'order' are judged by the human mind, this seems hardly relevant. What *is* relevant is that a gas which was previously believed to be a single substance is found to be capable of being separated into two or more components. Thus, when he says, 'the idea of the dissipation of energy depends on the extent of our knowledge', it would have been much more to the point if he had expressed the truism: 'whether or not a gas is known to be a pure substance depends on the extent of our knowledge.' When seen in this light the question of now attributing an entropy of mixing 'where previously we attributed none' is seen to be no more and no less a matter of human powers than is the knowledge of the purity of a chemical substance.

Leaving Maxwell, let us turn to Gibbs, some of whose remarks are closely similar in their import. For instance, in his comments on the probabilistic foundations of statistical mechanics, he said that probability refers to something which is 'imperfectly known'. He frequently referred to 'properties sensible to human faculties'; for instance, 'the grand ensemble would not differ to human faculties of perception from a petit ensemble....' And again he spoke of thermodynamic experiments as being made on bodies whose states 'are certainly not known to us exactly'. Perhaps this subjectivist implication was not fully intended by Gibbs, for he must surely have appreciated that almost *all* experiments, and not solely those of thermodynamics, are carried out on bodies whose states are not exactly known. But once more, in his discussion on the dispersion of a band of colour in a stirred liquid, which led on to the notion of coarse graining in statistical mechanics, Gibbs pointed out that, although the colour appears to become more uniformly spread, this is only the case if *our* considerations relate to samples of fluid whose thickness is much greater than the thickness of the colour filaments. (Gibbs, 1902; the quotations are from pages 17, 163, 203 and 145 respectively, of the Dover edition.)

Very similar views were later expressed by Tolman (1938) and G. N. Lewis (1930). The latter anticipated the ideas which were soon to arrive from information theory and wrote as follows:

> The increase in entropy comes when a *known* distribution goes over into an *unknown* distribution .... Gain in entropy always means loss of information, and nothing more. It is a subjective concept, but we can express it in its least subjective form, as follows. If, on a page, we read the description of a physico–chemical system, together with certain data which help to specify the system, the entropy of the system is determined by these specifications. If any of the essential data are erased, the entropy becomes greater; if any essential data are added the entropy becomes less.

This argument, which is a very important one for the subjectivist case, will be examined more carefully in §§2.5, 3.3 and in Chapter 5.

Later expressions of the same general viewpoint are to be found in the writings of Born, Rosenfeld, Reichenbach, Brillouin and others. For example, Born (1949) wrote that 'irreversibility is a consequence of the explicit introduction of ignorance into the fundamental laws.' This is a rather odd way of putting the matter, since no one, as far as we know, has literally introduced ignorance into the laws. No doubt Born was writing in a somewhat figurative or rhetorical sense, and he may have meant that classical statistical mechanics proceeds without using all of the available knowledge which *might* be used, i.e. by solving the differential equations for all of the $N$ molecules present in a macroscopic system. Even so, it can hardly be said that irreversibility is a *consequence* of this state of affairs: for even if it is true that entropy contains subjective elements, the effective irreversibility of such phenomena as temperature equalisation or nuclear synthesis in the stars is surely quite independent of our solving, or not solving, vast numbers of differential equations.

Rosenfeld (1953), too, made a strange remark: 'The irreversibility arises from . . . the incompleteness of our knowledge of the initial conditions . . . . The fact that the initial conditions of the system are known imperfectly . . . defines the thermodynamical aspect of the behaviour of the system.' Concerning thermodynamic *behaviour*, how could it be said, for instance, that the evaporation of a sample of water is 'defined' by imperfect knowledge? And on the question of how irreversibility arises, the same comment can be made as that already made about Born's contention.

Perhaps these statements and contentions were really nothing more than casual remarks on the part of their authors, believing that what they said and wrote was firmly established, requiring no further effort of mind. Some of the more carefully considered opinions of Grad, Jaynes and Hobson will be quoted later.

## 1.2. Objectivity and subjectivity

As has been seen particularly clearly in the quotation from G. N. Lewis, entropy was regarded as subjective precisely because it was taken to be a measure of incompleteness of knowledge. The belief that entropy is such a measure is, of course, one of the foundations of modern 'information theory' as developed by Shannon, Jaynes and others. But let us consider the notions of objectivity and subjectivity in rather general terms before taking up the issue concerning entropy.

To say that some physical phenomenon or entity is objective can be understood in two different senses, one of them weak and the other strong. The weaker meaning is that the occurrence of the phenomenon, or the existence of the entity, can be *publicly agreed*. Such would be the case, for instance, if it were said that ice is cold to the touch. Everyone can agree that this is so, but of course the assertion depends on the existence of the human nervous system. By contrast, the strong meaning of objectivity, the meaning most commonly used in science, is that the phenomenon or entity in question has a reality *quite independent of man's presence in the world*. This form of objectivity would apply, for example, to the event of the laying down of the Devonian rocks. Geologists would hold that the fully objective occurrence of that event is in no way impugned by man's non-existence at that period.

These two forms of objectivity (which may be called objectivity$_1$ and objectivity$_2$, respectively) are, of course, by no means mutually exclusive. Most phenomena or entities (or events or states-of-affairs) which fall into the category objective$_2$ can also be publicly agreed, and many of the things which can be publicly agreed may enjoy the status of objectivity in the stronger sense. As for the meaning of 'subjectivity', we shall simply take this as being the negation of one or the other of the meanings of objectivity.

Perhaps it may be objected that the foregoing is tacitly to adopt a 'realist' position in the philosophy of science – or at least to the extent of maintaining that there *is* some reality which is independent of mind, and hence of man's existence, and about it statements can be made which are either true or false.

To give adequate expression to the objections to realism would clearly require a long digression – one which would start in Plato's cave, go on to the views of Hume and Kant, and finally arrive at the controversies of the present day. From Kant we have the view that the world must be understood in terms of concepts, such as space and time, which are created by the human mind and are determined by its structure. We cannot be certain that these concepts really match the unknown 'things-in-themselves' to which we have no direct access. Feyerabend and others of the modern period have given a further slant to the Kantian criticism. All scientific work, they have said, even such a simple operation as the reading of an ammeter, must already be 'theory-laden' if it is to be meaningful. Meanwhile, from the viewpoint of the historian, Kuhn has drawn attention to certain rather abrupt and periodic changes in the very concepts used in science, thus leading to further doubt about whether there can ever be any truth in what science claims.

Our own position is close to Worrall's (1982) *conjectural realism*, i.e. to the view that our theories are best regarded as *attempts* – attempts truly to describe the world's contents and structure. Although theories change, our present theories are essentially the best conjectures we can presently make about the truth.

This is a modest form of realism and perhaps will not give full satisfaction to the practising scientist who finds much of his motivation in the feeling that he is uncovering a 'hidden reality'. But suppose one were to go in the opposite direction and adopt some form of anti-realism? How would our distinction between objectivity$_1$ and objectivity$_2$ look from that viewpoint?

Surely it would remain useful to distinguish those conjectures and factual statements which the scientist believes would still apply if man were absent from the world from those other conjectures and factual statements which are clearly dependent on the existence of humans? This distinction will continue to be useful, we suggest, whatever may be the eventual outcome of the philosophical discussion on realism versus conventionalism and relativism, positivism and instrumentalism.

Let us then adopt the distinction between strong and weak forms of objectivity, and adopt also a manner of speaking which will be consistent with conjectural realism.

However, it has still to be argued whether or not the *properties* of bodies, and more especially the *measurements* of those properties, can be objective$_2$. This involves two related issues, the first concerning scientific instruments.

Now it is true, of course, that instruments do not occur naturally; they are human artefacts. Yet it would be entirely contrary to the realist view of science if the measured values of physical properties were regarded as being dependent on man's existence. Conceptually, at least, we can think of measuring instruments, along with recorders, cine-cameras, computers, etc., as being present in the world even if man were not also present. Similarly, these instruments can be regarded as delivering 'readings' even if no humans take these readings into their thoughts.[1] (Superscript numerals refer to notes at the end of the book.) On the other hand, *inexact* or *erroneous* measurements can hardly be regarded as reflecting an objectively existing state-of-affairs! And how can any measurements ever be perfectly exact?

This leads on to the second issue already referred to, namely the status of the propositions which the scientist puts forward on the basis of his measurements; that is to say, his 'knowledge' or 'information'.

Since propositions are humanly created they cannot have the status of objectivity in the strong sense. Thus Popper's contention that 'knowledge is objective' amounts, in our view, to the claim that knowledge, or at least certain parts of it, has a fully objective *reference*, i.e. that it is *about* things or events which exist or occur quite independently of man's own existence. Our science, the realist believes, says things which are admittedly conjectural but which nevertheless refer to a world whose main features would be much the same if we were not there to make any conjectures at all.

If this viewpoint can be accepted we have already almost answered the question at issue: does incomplete knowledge about the physical world imply some element of subjectivity?

Whereas the exhaustive description of an abstract entity, such as an ellipse or a triangle, is always possible, the same obviously does not apply to physical entities and phenomena. These have a sort of inexhaustible quality and something would always be left unsaid. In place of the actually existing entities, or the actually occurring phenomena, in all of their natural complexity, we have to substitute a 'model'. It follows that knowledge in *all* fields of science (and not solely in thermodynamics and statistical mechanics) is inevitably incomplete and subject to revision. Is it not the case therefore that incompleteness of knowledge *cannot be a sufficient reason for imputing subjectivity*? If, as has been suggested, the objectivity of knowledge in science means that such knowledge has a fully objective *reference*, i.e. to real things or events, the amount of our information affects only the adequacy of our chosen 'model'. Even in situations where some additional or more exact measurements *might* be made this does not provide a reason for regarding our present results as being subjective. Indeed, if we had to wait for the attainment of entirely complete knowledge about the physical world before claiming that our knowledge has an objective reference, we should have to wait for an infinite time!

The matter can perhaps be clarified by discussing an example such as the example of the Devonian rocks. If we adopt the realist position, then both the occurrence of the laying down of these rocks and the date of that event are fully objective. Furthermore, if it were generally contended by geologists that the event probably occurred *n* million years BC, this contention would be an item of objective knowledge in Popper's sense. The word 'probably' indicates that the geologists recognise that they are dealing with an uncertain dating which is subject to revision. But this uncertainty and lack of complete information does not affect the objectivity of the event or its real date.

To be sure, the various learned geologists may be unable to agree among themselves. One of them might claim that the event occurred *m* million years BC, another that it occurred *n* million years BC, and so on. As contentions (as distinct from what is really the case), these would have a subjective character, either because the various experts have private access to particular items of evidence or because they assess publicly agreed items of evidence in different ways. However, an intrusion of subjectivity of this kind does not affect the main issue.

Let us briefly summarise. To accept that our knowledge is inevitably incomplete is one thing, but to claim that this incompleteness implies subjectivity is something quite different. In the application of any of our theoretical concepts, and in the consideration of any of our measurements, what we have to ask is whether the physical states-of-affairs to which these concepts or measurements refer can reasonably be supposed to exist independently of man. If so, our statements and propositions are *about* something which is fully objective and it would appear that the *amount* of our information can have no bearing on the matter. Thus lack of complete information is not, in general, a sufficient condition for an imputation of subjectivity; what has to be considered in this chapter, and in the following one, is whether the case of entropy may be an exception to this general rule, i.e. whether lack of complete information is here a sufficient condition, or whether it is but one of a number of other necessary conditions which, in conjunction, may be sufficient to render entropy subjective.

### 1.3. Thermodynamic entropy

In an important sense entropy *is* precisely what it is defined to be in classical thermodynamics since it was in this context that the term 'entropy' was first created by Clausius. To be sure, the notion of entropy has been broadened by the advent of statistical mechanics, and it has been still further broadened by the later advent of information theory. Nevertheless it remains to be seen whether these enlarged concepts of entropy are precisely the same as the originally defined thermodynamic entropy, and whether the alleged subjectivity applies to some of these concepts and not to others.

The original thermodynamic entropy was defined as a differential:

$$dS \equiv (dq/T)_{rev.} \tag{1.1}$$

That is to say, the infinitesimal increase of the entropy of a body in any *reversible* process which it undergoes is equal to its infinitesimal intake of heat divided by its absolute temperature. Thus the classical definition allows only of the obtaining of changes of entropy, $S_2 - S_1$, between states 1 and 2 when these are connectible by a reversible path, i.e. when there exists a path between states 1 and 2 which passes through a continuous sequence of states, none of which differ more than infinitesimally from an equilibrium state. This limits the thermodynamic definition of entropy differences to the differences between states of equilibrium.

It will be clear that such differences

$$S_2 - S_1 = \int_1^2 (dq/T)_{rev.} \tag{1.2}$$

are fully objective, i.e. are objective$_2$, *so long as* the heat quantities and temperatures along the reversible path (if it is available) are also objective$_2$. (Types of process where no reversible path is available are discussed by Bridgman, 1950.)

Consider first the heat. Normally it will be given to the body in question, which is at the temperature $T$, by a heat reservoir whose temperature must differ only infinitesimally from $T$. The heat given up by this reservoir can be restored to it, and measured, by performing mechanical work on the reservoir, e.g. by use of a weight and pulley. Thus heat can be expressed quite precisely in terms of the change of potential energy of a weight. In practical applications of (1.2) the aim is usually to use only one, or a few, such reservoirs, each of them large enough to remain at essentially constant temperature. Thereby (1.2) can be expressed as a sum, but that is a detail which need not concern us.

No doubt there are certain situations where a heat quantity may be *ambiguous*, though not necessarily subjective. For example, if a fluid is in turbulent motion it may be difficult to state precisely at any moment how much of the total kinetic energy is to be counted as thermal energy. However, this is a non-equilibrium situation and is not one with which classical thermodynamics can deal.

Turning now to the temperature, $T$, there can be little doubt that this, too, is fully objective and that there is little risk of ambiguity provided the process in question does not depart more than infinitesimally from equilibrium. For if this proviso is satisfied, the various degrees of freedom – translational, rotational and vibrational – of the molecules comprising the heat reservoir (which for simplicity can be taken as gaseous) remain at equilibrium with each other and the Maxwellian distribution will be maintained. And, of course, it is to those degrees of freedom, rather than to those which are internal to the nucleus, to which temperature measuring instruments are sensitive. It is only under rather extreme conditions, such as occur in the flame front of burning gases, that temperature measurement may become ambiguous due to the translational, rotational and vibrational 'temperatures' becoming distinct from each other (Meixner, 1941; Prigogine, 1949).

The same conclusion about the objectivity of thermodynamic entropy may be drawn from the familiar Gibbs' equation:[2]

$$dU = T\,dS - p\,dV + \sum \mu_i\,dn_i \qquad (1.3)$$

Here the variables determining a change $dS$ of entropy are $U$, $V$ and the $n_i$ whose differentials are the changes of internal energy, volume and mole numbers respectively. Can it be doubted that these are fully objective? If not, it follows that, under conditions such that the Gibbs' equation holds, the entropy change $dS$ is also fully objective.[3] There can be no difference in the objectivity status of the various variables in the equation.

Grad (1967) objected that 'there are several choices' for thermodynamic entropy. He instanced the case of a chemical reaction; if it is slow, he wrote, we have the choice of two distinct entropies which are functions of different variables. However, there is nothing here which is peculiar to entropy, since any other composition-dependent property of the reacting system would be similarly affected. We might 'idealise' the system by regarding the reaction as being so slow that its actual occurrence may be neglected, and thus all of the $dn_i$ in the Gibbs' equation would be taken as changing independently. Alternatively, we might idealise the system by regarding the reaction as sufficiently fast for it to be at reaction equilibrium, whereby not all of the $dn_i$ are independent and those which are not may be eliminated. Yet idealisations of this kind are surely to be regarded as convenient *approximations*, of varying degrees of adequacy, and have little bearing on the issue of subjectivity.[4]

In his well-known paper, *The Many Faces of Entropy*, Grad (1961) also remarked that a change in the estimated value of an entropy can occur 'when some relevant facet of the problem at hand has changed, even if only in the mind of the observer'. A familiar example is provided by Third Law entropies: the scientist may have to change his tabulated values as soon as he learns that the substances in question contain previously unknown isotopes or that the crystals of these substances display a residual randomness at the lowest attainable temperatures. Similarly, he may do so for reasons concerning nuclear spin. However, the fact that an entropy value is changed or is not changed, according to whether it is known or not known that such factors are operative, is no more an indication of subjectivity than would be the change of almost any other numerical value. Consider our previous example of the laying down of the Devonian rocks; the approximate date of this event is not normally regarded as being subjective simply because our knowledge of it is liable to revision. There is no good reason why our present estimates of Third Law entropies should not be regarded as being approximations to an objective$_2$ reality, just as we regard our present estimates of the age of rock strata as also being approximations to an objective$_2$ reality.

Some further remarks in answer to Grad's point will be useful at this stage. Now it is true, of course, that the use of Third Law entropies requires the consistent application of certain conventions. In particular, any entropy within the nucleus must be taken as zero, if only because it is entirely beyond calculation at the present state of theory. What saves the situation, as far as chemistry is concerned, is that the unknown nuclear entropies can be assumed to remain constant during the occurrence of chemical processes for which one requires to calculate only *changes* of entropy. The same applies to the entropy of isotope mixing, for this, too, will cancel out –

except, of course, in processes where there is a partial, or complete, separation or a mixing, of the isotopes.[5]

A strong argument *against* the charge of subjectivity is the normally very satisfactory agreement which obtains between Third Law calorimetric entropies on the one hand and spectroscopic entropies on the other, when all well-understood factors have been allowed for. Our point is that calorimetric and spectroscopic entropies are calculated by use of *two virtually independent theories*. The former are based on classical thermodynamics and on the integration of $dq/T$ over the whole range from a temperature close to the absolute zero up to the reference temperature (298.15 K) at which the entropy values of substances are tabulated. The spectroscopic entropies, on the other hand, are based on the theory of statistical mechanics together with the use of spectroscopic data for the purpose of evaluating the partition function of the particular substance in its ideal gaseous state at the reference temperature.

Aston (1942) tabulated data on 22 substances and these show that discrepancies greater than 0.6 per cent occur in only four instances – namely $CO$, $N_2O$, $H_2$ and $H_2O$. However, these four larger discrepancies can be accounted for very reasonably; in the case of $CO$ and $N_2O$ by assuming that with these linear molecules there occurs an end-to-end frozen-in randomness in the crystal at the lowest temperatures, requiring a correction to the calorimetric entropy of $R \ln 2$; in the case of $H_2$ because of the co-existence of ortho- and para-hydrogen in the ratio $3:1$; and in the case of $H_2O$ due to a type of randomness in the crystal requiring a correction of $R \ln 6/4$. With these acceptable corrections the difference between the calorimetric and the spectroscopic entropies for all 22 substances was less than 0.6 per cent – and may have been reduced still further subsequent to the publication of Aston's compilation.

In short, the experimental evidence indicates that the entropies calculated by use of two quite distinct theories are in close agreement and no doubt can be brought into even closer agreement by use of more refined techniques. This kind of numerical convergence thus supports the view that the calculated entropies correspond to an objective reality, even though this can only be approached and perhaps never be fully attained.

Leaving Grad's point, let us move on to one made by Jaynes (1965). He began by remarking that a given *physical* system can correspond to many different *thermodynamic* systems, according to the choice of the variables which are used for specifying its state. This is true and is fully in line with the views already expressed. However, Jaynes went on to say, 'From this we see that entropy is an anthropomorphic concept, not only in the well-known statistical sense that it measures the extent of human ignorance as to the micro-state. *Even at the phenomenological level, entropy is an anthropomorphic concept.* For it is a property, not of the physical system, but of the particular experiments you or I choose to perform on it.' (The italics are his.)

This last sentence, which is used to entail the previous one, is remarkably tendentious and could equally well have been written omitting the 'you or I', as follows: 'It is a property of the variables required to specify the physical system under the conditions of the particular experiment.' Jaynes proceeded to fortify his charge of anthropomorphism by making reference to the engineer's steam tables. Since steam is a polar substance, he said, its entropy 'depends appreciably on the electric field strength present. It must always be understood implicitly ... that the electric field was not inadvertently varied from one measurement to the next.'

This, of course, is true, but there is nothing here which is peculiar to entropy! The values of many other 'properties of steam' (to use conventional scientific language) will also be dependent on the field strength, and will be ambiguous if the field is not specified, along with the other state variables.

It may be remarked, more generally, that it is important to distinguish between the objective reality of some physical property, on the one hand, and the *value* attributed to that property on the other. The latter depends on *our* state of knowledge in a way in which the existence of the property itself does not. Some of the charges of subjectivity which have been brought against entropy may owe much of their apparent substance to a failure to draw this distinction.

Rather than accept that entropy is an anthropomorphic concept, the present authors prefer the way in which the matter has been expressed by Hobson (1971), whose views in other respects are close to those of Jaynes. Instead of speaking of the entropy of the physical system, Hobson thinks we should speak of 'the entropy of the data'. He also remarks that this does not render entropy subjective for it is an observable quantity and its value is determined by the observable data such as energy, volume and composition. The value is thus relative, not to the observer but to the observer's data. This way of putting the matter takes good care of our own points about isotopes, etc., for the data in question might, or might not, include information about the presence of isotopes as part of the system's composition variables. It will be clear, however, that Hobson's position is a very general one and does not solely concern entropy; *any* physical property is a function of all variables which determine the state of the body in question, rather than being 'a property of the body' *simpliciter*.

This is a suitable place at which to comment on G. N. Lewis's remarks as quoted in §1.1: 'Gain in entropy always means loss of information, and nothing more .... If any of the essential data are erased, the entropy becomes greater; if any essential data are added the entropy becomes less.' There is one sense in which this statement is correct and another in which it is seriously misleading or even erroneous.

It is true that, when we are concerned with certain aspects of statistical mechanics, the use of less information than may be available does, indeed, result in a greater calculated entropy. Thus a coarse grained entropy increases in time and becomes greater than the original fine grained entropy of the same system. Perhaps it was this which Lewis had in mind.

On the other hand, there are circumstances in which his statement is actually false. For example, the *gaining* of information that a spectral line, previously believed to be a singlet, is actually a multiplet, will require a *raising* of the spectroscopic entropy. The same applies to the gaining of knowledge that our system contains previously unsuspected isotopes so that we may now allow for an entropy of isotope mixing. A similar situation prevails in the four instances mentioned earlier where the calorimetric entropy was significantly less than the spectroscopic entropy. In these instances the value of the former entropy was *raised* by using information, or data, about the presence of randomness in the system – or in the case of $H_2$ by using information about the co-existence of ortho- and para-hydrogen.

Lewis's remark needs therefore to be very considerably qualified.[6] It is certainly not true *in general* that lack of data, or of information or knowledge, necessarily implies a greater value of the entropy. It is far better to think in terms of *the correction of error*, in whichever direction it may occur, than to conceive of some entirely general or universal connection between 'information' and entropy. In short, we conclude that thermo-dynamic entropy is a fully objective characteristic of the data on a system, of how that system is specified, and in this respect it is entirely on a par with physico–chemical properties in general. Whether or not entropy itself is correctly regarded as a 'property' will be discussed in §2.6.

## 1.4. The increase of entropy

The foregoing has been concerned with the *existence* of entropy as a function of state. In classical thermodynamics this existence theorem, together with the definition of absolute temperature, is demonstrated by using either the Clausius or the Kelvin 'impossibility statement' which are based on empirical observation. The increasing property can be demonstrated in the same way, i.e. by showing that if the entropy change of an adiabatically isolated system were supposed to be *negative* this would be contrary to the Kelvin statement concerning the impossibility of taking heat from a reservoir and obtaining its complete conversion into work without leaving a change of state in some other body. These demonstrations are available in the textbooks.

However, there are a couple of observations concerning $dS/dt \geqslant 0$ which deserve mention. As noted already in connection with equations (1.1) and (1.2), the thermodynamic definition of entropy change is limited to reversible paths. The proposition that entropy increases in the spontaneous adiabatic change of a system from state 1 to state 2 can only be established in general if there exists *some alternative reversible* (and *ipso facto* non-adiabatic) path between the same two states. However, in certain types of process an alternative path of this kind may not be experimentally attainable. For example, it has not yet been possible to perform the transition between diamond and graphite in a reversible manner. Neither is it feasible to carry out reversible processes of radioactive decay or of nucleosynthesis. For this reason it is not strictly correct to regard the $\Delta S$ of these processes as being necessarily positive when they occur spontaneously under adiabatic conditions. But, of course, it would be extraordinary if in fact they failed to obey the Second Law. That is almost unthinkable! And, indeed, in some instances, e.g. the diamond/graphite transition, it is possible to confirm that $\Delta S$ is positive *without* the existence of a reversible path. This involves using the Third Law which allows the relative entropies of, say, diamond and graphite to be obtained at any desired temperature and pressure on the assumption that their entropies are equal at the absolute zero.

Our second observation is more important and relates to 'time's arrow'. Whenever it is asserted that entropy tends to 'increase', this is to presuppose a chosen direction of time. In particular, when we commonly write

$$dS/dt \geqslant 0$$

this is to take $+t$ as being towards 'the future' and $-t$ as being towards 'the past'. On the other hand, all of the most fundamental theories of physics are $t$-invariant and are thus unable to distinguish between the one direction of the $t$-coordinate and its opposite. Apart from the peculiar behaviour of neutral K-mesons there is no distinction, at the level of elementary particles, between 'past' and 'future'.

Of course, in human consciousness this distinction is clear: only the one direction along the $t$-coordinate appears as being accessible to the process of our 'living'. But this raises the question whether, when we speak of entropy as tending to 'increase', we are introducing a subjective factor.

That this is not so, at least from the standpoint of classical thermodynamics, can be seen from the fact that the Second Law is expressible in a form which makes no reference to a time-direction, or to an entropy 'increase'. Consider a number of macroscopic systems, $A, B, C$, etc., each of them adiabatically isolated. Let $t_i, t_j, t_k$, etc., refer to a sequence of clock readings, but without reference to which direction of the sequence is the direction of 'later than' or 'the future'. Further, let $S^A_{t_i}$, etc., be the measured entropy (relative to some standard state) of system $A$, etc., at the instant $t_i$.

Now the empirical content of the Second Law can be expressed as follows:

$$\text{If} \quad S^A_{t_i} \geqslant S^A_{t_j} \quad \text{then} \quad S^B_{t_i} \geqslant S^B_{t_j}$$

or

$$\text{if} \quad S^A_{t_i} \leqslant S^A_{t_j} \quad \text{then} \quad S^B_{t_i} \leqslant S^B_{t_j}$$

Therefore, in either case, we have:

$$(S^A_{t_i} - S^A_{t_j})(S^B_{t_i} - S^B_{t_j}) \geqslant 0 \tag{1.4}$$

since the brackets have *the same* sign whether this be positive or negative. A similar reasoning can then be applied to any other systems, $B, C, D$, etc., considered pairwise, and also to any pair of clock times.

The inequality (1.4) thus provides a statement of the Second Law which makes no reference to the temporal sequence. It is the universal parallelism of the entropy changes of the various adiabatically isolated systems which is the essential experimental result, and *not* that higher entropy states occur 'later'. To be sure we can, if we wish, take 'later' as being given by consciousness, but this subjective step is not necessary. Instead we can choose *any one* of the macroscopic systems, $A, B, C$, etc., as providing a temporal signpost for all the rest. For example, we could choose a particular sample of radium, carefully safeguarded in a standards institution, as providing an entirely objective definition of 'the future' according to recordings of its progressive 'decay'. The use of this system, say $A$, as a reference system for 'time's arrow' would not reduce the Second Law to a tautology since the parallelism of (1.4) would still apply as an *empirical fact* to *all* other systems, $B, C$, etc.

The foregoing represents, of course, a strictly classical viewpoint. As soon as one allows for the statistical character of the Second Law, and thereby for the existence of fluctuations, it can no longer be said with complete certainty that *all* macroscopic systems display entropy changes in parallel. There is the remote possibility that one or more of them may be undergoing fluctuations which result in entropy changes in the reverse sense to the rest. Nevertheless, it can still be asserted that the great majority of systems within a large ensemble will show entropy changes in parallel so long as they are 'initially' in far-from-equilibrium states.

But what does this proviso mean in view of the word 'initially'? In Reichenbach's account of the matter (1956), he accepted that the universe as a whole is in a non-equilibrium state with the consequence that small portions of it can be 'closed off', as isolated systems, and are thus 'initially' in non-equilibrium states. He called them 'branch systems'. The words 'initially' and 'closed off', which seem to presuppose a time direction, in fact merely represent the use of a convention which is chosen so that the time-direction defined by the entropy changes of the majority of the branch systems agrees with the time-direction as experienced in consciousness. Thus (1.4), although it is not specifically used in Reichenbach's analysis, continues to hold on the average.

The proposition 'entropy tends to increase' is thus seen to have no essentially subjective connotations. It can best be understood by recognising that the universe is an evolving system – 'evolving' in the sense that the large-scale features of its history, as we believe this history to be, do not display $t$-invariance. Indeed, it would not be possible for laboratory systems to be 'prepared' in non-equilibrium states but for the non-equilibrium state of the Sun which allows of the occurrence of photosynthesis and the existence of coal and oil deposits, etc. When these systems are followed along their world-lines they display parallel gradients of increasing entropy in the one direction of their world-lines and parallel gradients of diminishing entropy in the reverse direction. It is probably no accident that it is the time-direction of increasing entropy which we experience as being towards 'the future' since it is difficult to conceive how we could exist or survive under the conditions of what might be loosely termed 'anti-Second Law' phenomena.[7]

# 2

## *Is entropy subjective in statistical mechanics?*

### 2.1. Introduction

We have argued that thermodynamic entropy is not different, in regard to its status of objectivity, from physical properties in general. How came it then that so many scientists have held, and still hold, the opposite opinion? No doubt some part of the answer is due to the creation of statistical mechanics, for this showed that the Second Law cannot be an absolute law but is true only on the average, i.e. in a statistical sense. This conclusion was supported by the discovery of important fluctuation phenomena. Thermodynamics was unable to deal with these phenomena whereas statistical mechanics did so with great success. It was thus a natural development that the statistically defined entropies came to be regarded as more fundamental than thermodynamic entropy. And many scientists hold the view that the former involve a significant element of subjectivity.

There are, of course, quite a number of statistical mechanical entropies. Boltzmann defined three of them and Gibbs defined another four. They differ somewhat in their properties, and only the ones which depend on the convex function $x \ln x$ were used almost equivalently by the two pioneers. Subsequently, several more 'entropies' have been created, especially in information theory. Because some of them bear little relation to the original thermodynamic entropy, confusion can be reduced by adopting Gibbs' usage of the term 'entropy analogues' as applied to all of the statistical definitions.

The making of an adequate survey of the possible existence of subjective factors within statistical mechanics requires the provision of theory which will be familiar already to the physicist and chemist – although perhaps less so to the philosopher. So as to shorten the reading, the main text will concentrate on the assumptions of the theory and the formal parts will be given as appendices. For subjectivity, if it is present, is to be looked for in the assumptions and not in the mathematical development.

Only quantum statistical mechanics will be discussed since it is closer than classical statistical mechanics to the real physical situation. In this chapter we shall confine ourselves to the statistical mechanics of equilibria whereas the approach to equilibrium will be taken up in Chapter 3. All matters relating to the indistinguishability of particles are deferred to Chapter 4.

We begin with a broad outline of those aspects of quantum mechanics which are of particular relevance to statistical mechanics, paying particular attention to those aspects where subjective features might be suspected. In Appendix 2.1 the mathematical and physical axioms of quantum mechanics are put forward in greater detail, although without referring to the well-known experimental evidence on which they are based.

## 2.2. A survey of quantum mechanics

### 2.2.1. *States of a system*

In quantum mechanics, just as in classical mechanics, it is possible to do a series of measurements on a system which establishes its state. Yet quantum mechanics differs from classical mechanics in so far as the former, even though the state of the system has become known, is usually unable to predict the results of further measurements in advance; instead it provides only the probabilities of outcomes. It would be out of place to describe the historical developments which led to the replacement of classical mechanics by quantum mechanics in the atomic domain. It is sufficient to say that there appear to be no experimental results which are contrary to quantum mechanics although there are many which are contrary to classical mechanics.

The 'classical' state of a system of $N$ particles is determined by the position and momentum of each of them. Thus $6N$ variables have to be known and the state of the system can be thought of as a point in a $6N$ dimensional space – the 'phase space'. In classical mechanics the result of any further measurement at any time can then be predicted.

One of the well-known features of quantum mechanics is that the precise positions and momenta of the particles cannot be obtained simultaneously. Take the case of a single particle in one-dimensional motion in a box. A measurement $_M X$ will determine, to within a certain accuracy, the distance $X$ of the particle from the left-hand wall. Suppose this is immediately followed by a measurement $_M P$ of the particle's momentum. The effect of this measurement is that the previously measured position of the particle can no longer be attributed to it. In fact, the greater is the accuracy of $_M P$, the smaller is the degree of correlation of *a second* measurement of $X$ with the result of the first measurement.

Nevertheless, the quantum theory can provide *the probability* that a given measurement will yield a particular result. For example, an energy state, determined experimentally, can be expressed as a linear combination of position states and this provides the probabilities of the various position states.

What has just been said amounts to the important Superposition Principle. Given any two possible quantum states, $|\psi\rangle$ and $|\phi\rangle$, of the system and any two complex numbers, $a$ and $b$, then $a|\psi\rangle + b|\phi\rangle$ defines a unique new state which is also a possible quantum state of the system. More generally, any linear combination of states defines a further unique state. Because vectors have a similar property – any linear combination of vectors is another vector – it is therefore convenient to 'model' the quantum states by vectors in a vector field. Although the vector space allows vectors of all lengths, those vectors which correspond to physical states of a system should normally be chosen to be of unit length. For otherwise, as will be seen shortly, the sum of the probabilities of all of the system's quantum states would not be unity.

A vector corresponding to the state of a system is called a *state vector* and a vector not of unit length has to be multiplied by an appropriate factor in order to be eligible. Furthermore, if a state vector is multiplied by a complex number of modulus one, it continues to represent the same physical state. For the latter depends only on the direction of the vector and not on any multiplicative factor. For this reason it is often convenient to group all vectors of the form $c|\psi\rangle$, for fixed $|\psi\rangle$ and variable $c$, together into a *ray* and to call them a *state ray*..

The vector space turns out to have the properties of what is known as a Hilbert space and one can choose a set of vectors of unit length, called a *basis set*, which span this space. This set may comprise a finite or an infinite number of vectors, but it is always possible to arrange them in a sequence. Let the chosen basis vectors be denoted $|1\rangle, |2\rangle, |3\rangle$, etc. Any vector of the form

$$|\psi\rangle = \psi_1|1\rangle + \psi_2|2\rangle + \psi_3|3\rangle + \psi_4|4\rangle +, \quad \text{etc.} \tag{2.1}$$

belongs to the Hilbert space when the $\psi_j$ are any set of complex numbers such that $\sum_j |\psi_j|^2$ converges to a finite limit. The square of the length of $|\psi\rangle$ is $\sum_j |\psi_j|^2$.

Vectors belonging to the Hilbert space, denoted by symbols such as $|1\rangle$, $|\psi\rangle$, etc., are called *kets*. To each of them may be assigned a corresponding *bra* vector, such as $\langle\psi|$, in a bra sub-space. The kets and bras are related as follows:

$$\text{if} \quad |\psi\rangle = \sum_j \psi_j |j\rangle \quad \text{then} \quad \langle\psi| = \sum_j \psi_j^* \langle j|$$

where $\psi_j^*$ is the complex conjugate of $\psi_j$. It is part of the mathematical axioms that a bra vector $\langle\phi|$ can always be combined with a ket $|\psi\rangle$ to produce $\langle\phi|\psi\rangle$ which is a scalar quantity and is often called the *inner product* of $|\phi\rangle$ and $|\psi\rangle$. If $|\psi\rangle = \sum_j \psi_j|j\rangle$ and $|\phi\rangle = \sum_j \phi_i|i\rangle$, then

$$\langle\phi|\psi\rangle \equiv \sum_j \phi_j^* \psi_j \tag{2.2}$$

Two vectors whose inner product is zero are said to be *orthogonal* to each other. It is shown in Appendix 2.1 that if $i \neq j$ then $\langle i|j\rangle = 0$ and hence the basis vectors form an *orthonormal* set since they are of unit length as well as being orthogonal. $\langle\psi|\psi\rangle$ is another expression, equal to $\sum_j |\psi_j|^2$ (since $\langle j|j\rangle = 1$), for the square of the length of $|\psi\rangle$.

The inner product is used to express probabilities and this is its importance in statistical mechanics. Suppose that measurements reveal that the system is in state $|\psi\rangle$. The probability that an immediately following experiment will show the system to be in state $|\phi\rangle$ is $|\langle\phi|\psi\rangle|^2$. Consider the case of a particle in one-dimensional motion in a box. Denote by $|x\rangle$ the state in which the particle is exactly $x$ cm from the left-hand side of the box. If $\Delta$ is very small the probability that the particle lies between $x - \Delta/2$ and $x + \Delta/2$ is $|\langle x|\psi\rangle|^2\Delta$. Hence the probability that the particle lies between $x_1$ and $x_2$ is $\int_{x_1}^{x_2} |\langle x|\psi\rangle|^2 \, dx$.

$\langle x|\psi\rangle$ is a function of $x$ and is often written $\psi(x)$ and called the *wave function*. $\psi(x)$ can be calculated using the Schrödinger equation (Appendix 2.1). The wave function is a particular representation of the state function $|\psi\rangle$ called the Schrödinger representation and is very convenient for dealing with certain kinds of problem.

Although we have been writing, for purposes of illustration, about single particles, the postulates of quantum mechanics are assumed applicable to systems of any degree of complexity. This is of great importance in statistical mechanics where the systems in question may typically contain $10^{23}$ particles or more.

### 2.2.2. Observables

When some property $A$ can be measured by an experiment, $_MA$, that property is called an *observable*. Its value may be given by either a single number or by a set of numbers, as when the positions of one or more particles are measured in three dimensions.

In classical mechanics, as has been said, if the state of the system is already known, the value of $A$ can be calculated with certainty in advance of doing the measurement $_M A$. We can say $A = A(x, p)$ where $(x, p)$ is the point in phase space where the system lies. For example, if the system is a particle of mass $m$ moving in one dimension in a potential $V(x)$, the energy $H$ can be calculated, as $H = p^2/2m + V(x)$, in advance of the measurement of this observable, from the knowledge of $(x, p)$.

In quantum mechanics, by contrast, the state of the system is not expressible as a point in a finite-dimensional space, as in classical mechanics, but is a point in a Hilbert space. Even when the latter is of finite dimensions it still bears no resemblance to the phase space of classical mechanics. To any observable, $A$, there now corresponds an *operator*, $\hat{A}$, which acts on a ket vector to produce another ket vector in the same Hilbert space. Let the former ket be the state vector. $\hat{A}$ acts on $|\psi\rangle$ to produce another ket $|\phi\rangle$. This is expressed by the notation

$$\hat{A}|\psi\rangle = |\phi\rangle \tag{2.3}$$

so that

$$\langle\psi\,|\,\phi\rangle = \langle\psi|\hat{A}|\psi\rangle$$

As has also been said, it is not usually possible to predict with certainty the outcome of an experiment which measures $A$ even when the state $|\psi\rangle$ is known. However, if the experiment is repeated many times with the system in the same initial state $|\psi\rangle$ the average value of the result, known as the *expectation value* of $A$, is given by

$$\langle A\rangle = \langle\psi|\hat{A}|\psi\rangle \tag{2.4}$$

Since $\langle A\rangle$ is the average value of a physical property it must be expressed by a real number, not by a complex number. This requirement restricts the operators which represent observables to a special class known as the *hermitian operators*. Their full characterisation is discussed in Appendix 2.1.

### 2.2.3. *Eigenvalues and eigenvectors*

Given a hermitian operator, $\hat{A}$, one can set up the so-called *eigenvalue equation*:

$$\hat{A}|\psi\rangle = a|\psi\rangle \tag{2.5}$$

where $a$ is a scalar quantity and is a real number. The possible values of $a$ are called the *eigenvalues* of $A$ and are denoted $a_1, a_2$, etc. The corresponding vectors $|\psi\rangle$ are the *eigenvectors* (or *eigenstates*) of $\hat{A}$ and are denoted $|\alpha_1\rangle$, $|\alpha_2\rangle$, etc. Thus

$$\hat{A}|\alpha_j\rangle = a_j|\alpha_j\rangle \tag{2.6}$$

The eigenvectors are thus special kinds of vectors such that operation on them by the operator leaves them unchanged except for multiplication by the eigenvalue.

It can be shown that

$$\langle \alpha_i | \alpha_j \rangle = 0 \quad \text{if } a_i \neq a_j \tag{2.7}$$

and it is often possible to use the eigenvectors as an orthonormal basis set when they have been normalised to unit length.

If $|\alpha_j\rangle$ is an eigenvector of $\hat{A}$ then so also is $c|\alpha_j\rangle$ for any complex number $c$. If the *only* eigenvectors with eigenvalue $a_j$ are of the form $c|\alpha_j\rangle$ then $a_j$ is said to be non-degenerate. But if this is not the case, $a_j$ is called a *degenerate* eigenvalue. In general, if $|\alpha_{j_1}\rangle$ and $|\alpha_{j_2}\rangle$ are eigenvectors of $\hat{A}$ with eigenvalue $a_j$ then so is $b|\alpha_{j_1}\rangle + c|\alpha_{j_2}\rangle$ for any complex numbers $b$ and $c$. The set of all eigenvectors of $A$ with eigenvalue $a_j$ form a vector space called an *eigenspace* of $\hat{A}$. The dimensionality, $d$, of this eigenspace is the degeneracy of $a_j$. When $d$ is unity, $a_j$ is non-degenerate.

The eigenvalues and eigenvectors have an important physical significance in measurement. When the observable $A$ is measured on the system $S$ this implies that some macroscopic external system is influenced by the state of $S$, and reciprocally. The result of the measurement is the value of a reading taken from the external system – the instrument – or a quantity calculated from that reading. The result must be one of the eigenvalues of $\hat{A}$.

It was remarked above that probabilities are obtainable from inner products. Suppose that immediately *before* measurement of $A$ the system was in the state $|\psi\rangle$. If $a_j$ is a non-degenerate eigenvalue of $\hat{A}$ then the probability $p_j$ that the result of measurement will be $a_j$ is

$$p_j = |\langle \alpha_j | \psi \rangle|^2 \tag{2.8}$$

where $|\alpha_j\rangle$ is the eigenvector of $\hat{A}$ with eigenvalue $a_j$. On the other hand, if $a_j$ is degenerate, let the corresponding eigenspace, denoted $\alpha_j$, be spanned by the orthonormal set of eigenvectors $|\alpha_{j1}\rangle, \ldots, |\alpha_{jd}\rangle$ where the last term is numbered by the degeneracy $d$. Let

$$p_{ji} \equiv |\langle \alpha_{ji} | \psi \rangle|^2 \tag{2.9}$$

Then the probability that the result of the experiment is $a_j$ is given by

$$p_j = \sum_{i=1}^{d} p_{ji} \tag{2.10}$$

Conversely, if the result of the experiment is found to be $a_j$, then something can be inferred about the state of the system *after* measurement – namely that the state lies in a $d$-dimensional space, say $\alpha_j'$, spanned, say, by the $d$ vectors $|\alpha_{j1}'\rangle, \ldots, |\alpha_{jd}'\rangle$. Of course, if $a_j$ happens to be non-degenerate, there is only the one possibility and the state of the system after measurement can be inferred to be a particular eigenstate of $\hat{A}$.

If the time taken by measurement is infinitesimal, the $|\alpha'_{ji}\rangle$ are the same as the $|\alpha_{ji}\rangle$. However, any state $|\psi\rangle$ does not usually remain unchanged but evolves with time in the manner described below. This means that if the measurement is of finite duration the $|\alpha'_{ji}\rangle$ are not the same as the $|\alpha_{ji}\rangle$.

It should be added that the foregoing refers only to what are called *ideal* measurements (D'Espagnat, 1976), i.e. those which if repeated immediately on the same system give the same result. A typical *non*-ideal measurement (also said to be of the 'second kind') is the location of a particle by its absorption in a photographic emulsion. This is obviously not repeatable on the same particle, and involves an irreversible process.

### 2.2.4. *The Hamiltonian*

The Hamiltonian operator, $\hat{H}$ (whose construction is described in Appendix 2.1), has certain special properties which correspond to the fact that, in classical mechanics, energy is a constant of motion and does not change with time. Since the energy is obtained by evaluating the Hamiltonian, the latter is also a constant of motion. Furthermore, the eigenvalues of $\hat{H}$ are the system's 'energy levels' and if a quantised system is in an eigenstate of $\hat{H}$ it remains in that eigenstate for all time, or until the system is disturbed by some external force.

In the case of an enclosed system, another important property of the Hamiltonian is that its eigenvalues *are discrete*, so that there is always a gap between one energy level and the next, even though the eigenvalues may be degenerate. Thus the sequence of levels is unlike the set of real numbers which form a continuum.

The eigenvectors of $H$, denoted $|1\rangle, |2\rangle, \ldots, |j\rangle, \ldots$ (with corresponding eigenvalues $\varepsilon_1, \varepsilon_2$, etc.), may be used as a basis set. Thus any ket $|\psi\rangle$ can be expressed in the form of equation (2.1):

$$|\psi\rangle = \sum_j \psi_j |j\rangle$$

where the $|j\rangle$, as has been said, are independent of time in the case of an isolated system. On the other hand, the coefficient $\psi_j(0)$ becomes $\exp(-i\varepsilon_j t/\hbar)\psi_j(0)$ after a time $t$ and this is another peculiarity of the Hamiltonian when its eigenstates are taken as a basis set. Each $\psi_j$ can be thought of as rotating with angular velocity $\omega_j$ where $\omega_j = \varepsilon_j/\hbar$ ($\hbar$ is Planck's constant divided by $2\pi$). Thus after a time $t$ the ket $|\psi\rangle$ becomes $|\psi(t)\rangle$ where

$$|\psi(t)\rangle = \sum_j \psi_j(0) \exp(-i\varepsilon_j t/\hbar)|j\rangle \tag{2.11}$$

The probability $p_j(t)$ that the system is in state $|j\rangle$ at time $t$ is given by equation (2.8) and is $|\langle j|\psi(t)\rangle|^2$. Applying (2.11) we thus obtain:

$$p_j(t) = |\psi_j(0) \exp(-i\varepsilon_j t/\hbar)|^2 = |\psi_j(0)|^2 \qquad (2.12)$$

It follows that the probability that an energy measurement gives the value $\varepsilon_j$ does not vary with time so long as the system remains completely undisturbed. Furthermore, as shown in Appendix 2.1, the expectation value of the energy is

$$\langle \psi|\hat{H}|\psi\rangle = \sum_j p_j \varepsilon_j \qquad (2.13)$$

and this, too, is time invariant.

By contrast, consider some other observable $A$. Using (2.11), the probability that at time $t$ the system is in an eigenstate $|\alpha_\mu\rangle$ of the observable is

$$|\langle \alpha_\mu|\psi(t)\rangle|^2 = \left|\sum_j \psi_j(0) \exp\left(\frac{-i\varepsilon_j t}{\hbar}\right)\langle \alpha_\mu|j\rangle\right|^2$$

After expanding this expression, one obtains a sum of many terms whose phase angles rotate at different rates. Consequently the probability that a measurement of $A$ gives some particular value is not time invariant, and neither is the expectation value. It is because the energy states are unique in this respect that they are known as *stationary states*.

### 2.2.5. *Review*

Before proceeding to the statistical mechanics it may be useful to review the significance of the foregoing in the general context of this book.

As is well known, some theoreticians regard the state vector as being a fully objective feature of physical reality whereas others contend that it describes nothing more than a state of human knowledge. These controversies have been particularly lively in regard to the possible role of consciousness in the quantum measurement problem. As has been said, the process of measuring an observable normally involves a physical interaction between the system in question and a suitable measuring instrument. Since the reading of the instrument by the observer also involves an interaction, a number of quantum theorists hold that the observer himself must be counted as being a part of the measuring 'set-up'. In short, they maintain that the knowing subject is not fully separable from what he observes.

This subjective view is given some support by the fact that state vectors are not unique; the multiplication of a vector by a complex number has no physical consequences. Yet this point is of no significance in the application of quantum mechanics to statistical mechanics. For here it is not the state vectors themselves which are used but only the *probabilities* of eigenstates, together with the eigenvalues. As has been shown, these probabilities are given by the moduli of inner products, and the latter are indeed unique and are given by real numbers. So, too, are the eigenvalues, such as the energy eigenvalues $\varepsilon_j$, and their degeneracies. These have fully determinate values – although this is not to say that they are not dependent on a satisfactory choice of the 'model' for the system in question, such as is involved in the working out of its Hamiltonian. Using such a model, the theoretically calculated values of the $\varepsilon_j$ agree very closely with spectroscopic data and this gives strong support to the view that quantum mechanics is an expression of physical reality.

In short, the probabilities, together with the data which (as will be shown) are needed for evaluating the partition function, have a fully objective status in quantum theory. Furthermore, any entropy increase arising from perturbation during measurement on a system appears to be usually negligible in comparison with entropy changes which accompany macroscopic processes within the system itself. Indeed, if this were not the case, it could hardly be expected that there would be such a close agreement as normally prevails between the results of classical statistical mechanics and the results of quantum statistical mechanics, so long as due regard is given to the indistinguishability of like particles.

In the following, very little direct and quantitative use can be made of quantum theory except in the case of very simple systems such as perfect monatomic gases. The main purpose of the foregoing has been simply to provide a sound *conceptual basis* for statistical mechanics. To be sure, certain matters, discussed in detail by Ludwig and his school (e.g. Melscheimer, 1982), have not yet been touched on and these include the important differences between the notions of 'state' as used in quantum mechanics and in thermodynamics respectively. We shall come to these shortly.

Although there is intense controversy about the philosophical foundations of quantum mechanics (e.g. d'Espagnat, 1976; Mehlberg, 1980; Popper, 1982) our view in summary is that the eventual outcome will not significantly affect statistical mechanics. What is important for present purposes is that, whereas in classical statistical mechanics the statistics arose from the practical impossibility of having data on the positions and momenta of, say, $10^{23}$ particles, in quantum mechanics it is also a matter of basic principle that the results of measurements on a system are normally scattered probabilistically over a range of values. The statistical aspects of quantum statistical mechanics thus have a double origin: (a) in the essential incompleteness of the thermodynamic type of specification of a system, and (b) in the axiomatic structure of quantum mechanics itself. The former implies that the state vector is not part of the available data; the system in question is said to be in a 'mixed' and not in a 'pure' state.

### 2.3. The assumptions of equilibrium statistical mechanics

The following is a mere outline, but one which should be sufficient to cover those aspects of the foundations of the subject which are relevant to our general theme. As has been said, this chapter is limited to the discussion of equilibrium states – the same limitation as applies to thermodynamic entropy – and the consideration of irreversibility is deferred to the next chapter.

A useful beginning is the distinction between *macro-states* and *micro-states*. Thermodynamics is usually concerned with systems containing a very large number of molecules and specified in regard to having, say, fixed energy, volume and chemical composition, or fixed temperature, volume and chemical composition. Such a description refers to a macro-state (thermodynamic state) and would be vastly less detailed than the description of a micro-state (i.e. a quantum state) of the same large system. Thus an immensely large number of micro-states are comprised within the specification of any macro-state. And of course the micro-states in question refer to the quantum states of *the whole* macroscopic system, which is regarded as a single quantum entity, and not to the quantum states of the individual particles of which the system is composed. Only in the case of very dilute gases is it reasonable to regard a macro-system as being an assembly of independent particles, as was done in the original statistical mechanics of Boltzmann. In the case of other kinds of system, it is at best only an approximation to attribute to the component particles their own 'private' energy states, to use Schrödinger's felicitous phrase. In general, the quantum ideas must be applied to the system as a whole.

Particularly important for our purposes are the energy eigenstates and eigenvalues. For one reason because, as has been said, these eigenstates do not change with time in the case of isolated systems. For another, because of the mathematical simplicity of the canonical distribution, to be arrived at shortly, together with the additivity of energies for non-interacting, or weakly interacting, macro-systems. For a third, because the energy eigenvalues are *discrete*, in the case of enclosed systems composed of stable atomic or molecular species.

Following these preliminaries we turn to the two main methods of approach to the equilibrium statistical mechanics as used in the well-established treatises on the subject. These are the ergodic method and the method based on the postulate of equal *a priori* probabilities and random phases. Although the latter will be our choice for detailed discussion, it will be as well to make a few remarks about the ergodic method since it has certain features in common with the other.

The intention behind the ergodic method, which has been mainly developed in the context of classical mechanics, is to found statistical mechanics solely on the molecular dynamics of an isolated system. For this purpose one uses the *ergodic theorem* according to which the average value over *an infinite time* of a dynamical variable of the system in question is equal to the average value of this variable over the constant energy surface in the classical phase space. Certain sets of initial states of zero measure must be excluded, and also there must be no 'isolating constants of the motion' other than the Hamiltonian itself or functions of the Hamiltonian. For if there were such constants the 'representative point' (i.e. the point in the $6N$-dimensional phase space which represents the coordinates and momenta of the $N$ particles in the system) would be confined to some particular region of the energy surface. Still another condition is necessary if the system is indeed to reach a state of equilibrium, rather than permanently oscillating, and this is the so-called 'mixing' condition which is concerned with the loss of correlations. These various restrictions on the range of validity of the ergodic theorem have been shown to be rigorously satisfied only in certain instances – notably in the example, studied by Sinai, of a system consisting of hard spheres contained in a box with parallel sides which are perfectly elastic. Under other circumstances the ergodic method, in its present state of development, depends on probabilistic assumptions (Farquhar, 1964; Jancel, 1969; Lavis, 1977; Penrose, 1979). It would be tempting, wrote Lavis, to suppose that the method will remove 'the statistics from statistical mechanics', but he and others have shown that this objective is still far from being attained. The method has other drawbacks; it is difficult to extend it to the quantum situation and also, arising from the fact that it is concerned with infinite time averages, it has little to say about the approach to equilibrium or about fluctuation phenomena.

In the alternative method, as developed in the classical context by Gibbs and later adapted to the quantum situation by Fowler, Tolman and others, it is assumed from the outset that statistical mechanics is about systems which are *incompletely specified*, i.e. in the sense that the *required theory* is one which will apply to systems which are specified in regard to their macroscopic variables only. Since the state vector $|\psi\rangle$ is not a part of such a specification, the $p_j$ of equation (2.8) are not calculable. Therefore we require, as Tolman (1938) put it, an *extension of theory*, one which involves 'the introduction of an additional postulate not derivable from those included in the mechanics of precise states'.

Evidently, the required postulate is about the relative probabilities (or the weighting, to use Fowler's preferred term) of the immense number of micro-states comprised within the specified macro-state. As was said at the end of §2.2, even in classical statistical mechanics it was accepted that there is a practical impossibility of specifying the positions and momenta of some large number of particles in a container. The need for the postulate is even clearer in quantum statistical mechanics since, even if $|\psi\rangle$ were known, the quantum axioms would offer nothing more than a probability that the system in question will be found to have a particular value of one of its observables.

These points have contributed to what may be called the 'ignorance view' of entropy, as put forward by Jaynes and others. He remarks (1965) that entropy can be said in a subjective sense to measure 'our *degree of ignorance* as to the true unknown microstate, . . .'. And again (1957), '. . . it is not the physical process that is irreversible, but rather our ability to follow it.' But within his papers there is a sort of dichotomy. For elsewhere he accepts the objective reality of such phenomena as temperature equalisation, and also, in a very interesting passage (1957, p. 627) he remarks that statistical mechanics uses *subjective* probabilities for purposes of prediction, but should nevertheless be understood *objectively* for purposes of interpretation.

In our view it is quite irrelevant to call attention to human ignorance. This is because the *basic objective* of statistical mechanics is to create a theory whose final equations refer only to the macro-variables of a system – notably energy (or temperature), volume and chemical composition. In short, the incompleteness of specification is not dictated by what can be *known to us*; it is dictated only by the limits set for the required statistical mechanical theory. It has to be a theory which will underpin thermo- dynamics and which will refer at its end state (although not necessarily at its intermediate stages) solely to the macroscopic thermodynamic variables.

This is surely a more correct viewpoint than that of regarding statistical mechanics as providing a measure of *our* ignorance, or of rejecting information which might have been obtained if a maximal quantal measurement could have been made on the system.

Where we should look for subjectivity, if it is present at all, is in the basic postulate. As already said, its purpose is to bypass the need for data about the micro-states and it is as follows:

> A substantially isolated system whose energy lies in a narrow range $E$ to $E + \delta E$ has the property, when it is at equilibrium, that all of its accessible energy eigenstates within that range have equal probability; energy eigenstates outside that range have zero probability.

By 'equilibrium', we mean, of course, that the system has reached a condition in which its thermodynamic variables are constant in time.

Before considering the nature of this postulate a few technical comments should be made. The first is that the total energy of a quantum system, even though this energy is a macro-variable, cannot be exactly specified. This is due to the so-called energy–time uncertainty relation which states that when the energy of a system (relative to some chosen standard state[8]) is measured, or regarded as fixed, during a temporal interval $\Delta t$, the energy is uncertain by the amount $\Delta E \geqslant \hbar/2\Delta t$. The range $\delta E$, although narrow, must thus exceed $\hbar/2\Delta t$; but, of course, the latter can be regarded as almost vanishingly small in the case of a system at equilibrium since $\Delta t$ can then be taken as being indefinitely large.

A further comment is about an additional assumption of *random phases*. In equation (2.1) the expansion coefficients $\psi_j$ (also known as probability amplitudes) are complex numbers and can therefore be written $\psi_j = r_j \exp(i\phi_j)$. Here $r_j$ is a real number and $\phi_j$ is called the phase, or phase angle, of $\psi_j$. It is part of the basic postulate to suppose that the $\phi_j$ are random when the system is at equilibrium – that is to say, the equilibrium 'mixed' state is an incoherent superposition of energy eigenstates. Such an assumption depends on the system not being perfectly isolated – it must interact, even if only very slightly, with the external world – since otherwise the assumption would be contrary to the Schrödinger equation (equation (4) of Appendix 2.1).[9]

Where, as in the present chapter, we are concerned only with energy eigenstates, the effect of this additional part of the basic postulate is to simplify the density matrix and the concept of the density operator need not be introduced. In the 'representation' in terms of energy eigenstates, the density matrix is diagonal and its elements are equal to the probabilities $p_j$ of the occurrence of particular energy eigenstates in the 'mixed' quantum system at equilibrium.

Leaving aside these technical points, let us consider the basic postulate in relation to the subjectivity/objectivity issue. We ask, to begin with, whether the probabilities in question are subjective probabilities, or whether they come within the scope of the objective interpretations of probability, such as the frequency or propensity interpretations. Beyond that we ask whether the postulate itself can be justified on objective grounds.

The distinction between subjective and objective interpretations of the same probability calculus has been clearly drawn by Popper (1983, pp. 281, 295) and others. The subjective interpretation regards probabilities as expressions of human ignorance; due to our insufficiency of knowledge, and, even though the outcomes of events may be fully determined in advance, we can at best have only some 'reasonable degree of belief' about what those outcomes will be. The objective interpretation, by contrast, takes certain probabilities as being manifestations of real dispositional properties of physical systems, properties which are independent of man's presence.

These interpretations are not necessarily contradictory, for there are situations where the one may be more appropriate or more reasonable than the other. Take, for instance, a horse race – a unique and unrepeatable event. The expectation that a named horse will win naturally varies from one spectator to another – even though they use such objective evidence as is available to them about the states of health and previous records of the horses. The probabilities the spectators assign to the various horses winning, as measured by their betting quotients, can be regarded as subjective.

The situation is very different in statistical mechanics, or in the consideration of coin tossing or of radioactive decay. Here we are concerned with physical systems which can be specified, although not completely, but more importantly can be subjected to repeated 'trials'.

This allows the calculation of long-run frequencies. When, as in coin tossing, different people obtain close agreement on the relative frequency of heads and tails it seems reasonable to suppose that the frequency is an objective characteristic of the coin and of the tossing procedure.

Against this view of the matter it is often argued that it is logically inconsistent to assume that probabilities are objective when they refer to processes which occur according to fully deterministic laws. The variable behaviour as between one toss and another would then be a matter only of *our* incomplete knowledge about all the minute 'causes' which affect the outcome. For, if we knew the precise force exerted in the tossing, together with full details about air currents, etc., we could reliably predict each fall of the coin by use of the classical Newtonian mechanics.

We shall not discuss this argument in favour of subjective probabilities because in statistical mechanics we are concerned with systems which do *not* behave deterministically. *The best* that can be achieved in advance of the measurement of a quantum observable is the calculation of the probabilities, $p_j$, of the system being found in some particular eigenstate. These probabilities are given, as in equations (2.8)–(2.10), by the moduli of inner products and are not altered by any arbitrary change in the choice of the basis set. The $p_j$ are thus objective$_1$ in so far as all scientists who accept the axiomatic structure of quantum mechanics accept their existence. They are also objective$_2$ to the extent that quantum mechanics is an approximately true account of external reality; the propositions which formally state the values of the probabilities are truth functionals within the body of the theoretical system. Of course, as was said at the end of §2.2, the statistical aspects have a double origin and the $p_j$ are not actually calculable on the basis of nothing more than a macroscopic specification of the initial state. They figure in statistical mechanics rather as theoretical entities which provide an adequate conceptual basis.

No doubt the objection may be raised that, in the absence of the act of measurement, quantum mechanics is actually a deterministic theory. Isolated systems, it will be said, develop in time in accordance with the Schrödinger equation whose integral was given as equation (2.11). If so, the systems traverse a determinate sequence of eigenstates and are in a pure state at any moment.

However, the Schrödinger equation holds only under entirely idealised conditions – those of perfect isolation. It is a very important fact that real systems *are never perfectly isolated*; they inevitably undergo small disturbances such as might be due to the penetration of the walls of their containers by cosmic rays, or to the effect of fluctuating external fields of force, or simply to the effect of the impact of external molecules on the outside walls of their containing vessels.

As will be seen later, this point about the effects of minute external perturbations is of great significance to our whole approach. To be sure, we do not regard the perturbations as being the *origin* of irreversibility (for processes such as mixing would be expected to occur even if perfect isolation could be attained); we think of them rather as eliminating any possibility of deterministic temporal development at the molecular level.

Let us illustrate the point with a calculation made by Borel (1928, p. 174). He found that if a mass of one gram happened to be displaced through a distance of one centimetre on a star at the distance of Sirius it would affect the magnitude of the gravitational field at the Earth's surface by a factor of only $10^{-100}$; even so, this would be sufficient to completely falsify the classically calculated trajectories of the molecules in a container on the Earth's surface within a period as short as $10^{-6}$ s. An even more dramatic example has been quoted by Berry (1978) and attributed by him to Chirikov: the presence or absence of an electron at a distance of $10^{10}$ light years would alter the gravitational force at the Earth by an amount sufficient to change the angles of molecular trajectories by as much as one radian after only 56 collisions. Although these were classical calculations, Zeh (1970) and Berry (1978) have similarly shown that quantum systems containing many particles (and thus having dense energy levels) display extreme sensitivity to minute disturbances.[10]

How these random influences might be allowed for will be returned to in § 3.5. For the moment it is sufficient to say that their existence discredits the basis of the claim about logical inconsistency as mentioned on p. 31.

So far we have been discussing, and dismissing, the subjective interpretation of the probabilities $p_j$ as applied to the typical statistical mechanical situation. Let us turn to the objective interpretations of the formal probability calculus. One of them is that probabilities are to be understood as *limiting relative frequencies*, as obtained approximately by the making of a large, but finite, number of repeated trials. The other is the *propensity* interpretation where probabilities are taken as being the 'propensities of single events to realise themselves in long runs'. (Popper, 1983, p. 400.)

The frequency interpretation was used by Gibbs in a manner which was not inconsistent with the subjective viewpoint as already discussed. At his time, before quantum mechanics had been developed, there was still a strong belief in the completely deterministic behaviour of mechanical systems. Accordingly, the justification to Gibbs for the introduction of probabilistic notions into statistical mechanics was the lack, *to us*, of a sufficiently detailed knowledge of complex systems which would have allowed of the application of exact mechanics. Furthermore, the frequency interpretation was regarded by many scientists at that period as the only interpretation suitable for science. Its difficulties, in so far as it requires postulates of convergence and randomness (Gillies, 1973), were not fully appreciated.

In its application to statistical mechanics, Gibbs adopted the notion of *ensembles* of like systems, differing from each other only in regard to their microstates. The probability of any one of the systems being in a particular microstate was identified with the fraction of the systems within the ensemble which are in that state. Or sometimes it was said that the probability in question is the probability that a system chosen at random from the ensemble will be in that state (Tolman, 1938, p. 47). More will be said later about this concept of the ensemble.

Accounts of propensity theory are given by Popper (1956, 1983), Hacking (1965), Mellor (1971, 1983) and Gillies (1973) and are criticised by Kyburg (1974). To most of these authors a probability is a measure of the tendency or disposition of a physical system, together with its environmental circumstances, to behave in a certain way. The propensity thus refers to a 'set-up', as Hacking calls it, and this may be either natural or contrived. Take the case of the tossing of a penny. The propensity of the tossing set-up for the coin to fall heads up will be affected by any bias in the coin, by the manner of tossing, by air currents, by the surface on which the coin falls, and so on. (If it were to fall on mud it might remain edge-up, thus altering the propensity for heads and tails.) Mellor has a rather different usage since he identifies the propensity solely with the bias (if any) of the coin, so that the latter, rather than the whole set-up, becomes the bearer of the propensity. On either view of the matter we are dealing with an objective property even though it is dispositional, i.e. in the sense (like the fragility of glass) that it may or may not become manifest. The propensity interpretation thus appears to be a good alternative to the frequency interpretation for the purpose of providing a conceptual vehicle for the understanding of the probabilities of the occurrence of eigenstates.

Turning now to the justification of the basic postulate, this is usually done by reference to Keynes' Principle of Indifference – previously often known as the Principle of Insufficient Reason.

It is well known that the application of this 'Principle' can lead to paradoxes and absurdities unless good care is taken.[11] Some of these difficulties arise when the 'events' in the probability space can be partitioned in alternative ways. Think, for example, of a set of a hundred cards labelled from 1 to 100. The likelihood of blindly drawing a card from either the class of even numbers or the class of odd numbers may reasonably be considered as equal. On the other hand, if the cards were partitioned into the classes of the primes and the non-primes, a condition of equal likelihood would no longer hold. A related difficulty, discussed by Keynes, occurs when the various alternative outcomes are 'divisible', i.e. when they can be broken down into sub-alternatives.

Kyburg (1970) suggested that the 'Principle' is most justifiable when the physical situation is such that there is the presence of symmetry, and more particularly the presence of symmetry which is relevant to the various possible outcomes – for instance, in the case of the six faces of a die which is not known to be loaded. In Popper's discussion of the matter (1983, p. 375) he accepts this suggestion as being a useful *conjecture* but he rejects the idea that any 'Principle' is involved. He acknowledges that hypothetical estimates of equal probabilities, suggested by symmetry considerations, play an important part in physics. Even so, such estimates continue to be conjectures, to be tested by experience, and they are capable of being falsified – for example, by frequency tests such as could be applied to a die in order to discover whether or not it is significantly biassed.

In view of these points it is best that the 'Principle' be used as nothing more than an heuristic device, one which is capable of agreeing in many instances with the objective physical situation but is also capable of misleading. In an interesting passage, Jaynes (1979, p. 27) quotes Maxwell's use of the 'Principle' in his gas kinetic theory when he supposed that every position within the circle of collision of the molecules is equally probable. Jaynes remarks that it is our intuition which tells us that the probability of impinging on any particular region should be taken as proportional to the area of that region and not to, say, the cube of the area or its logarithm. 'In other words, merely knowing the physical meaning of our parameters, already constitutes highly relevant prior information . . . .'

This point about intuition may suggest that the use of Keynes' Principle is a subjective matter, but this would not do justice to the significance of subsequent experimental corroboration. All scientists use their intuition for much of the time! In Maxwell's case the consequences of his theory (some of which were actually counter-intuitive such as the pressure and temperature dependence of gas viscosity) were shown to be capable of satisfying very stringent experimental tests.

Mellor, in an important argument (1971, Chapters 6 and 7), avoids using the Principle of Indifference by using instead the Principle of Connectivity. This principle (which is not well known to scientists but is tacitly accepted by them) asserts that two things never differ from each other *in one property only* – there must be at least one other property by which they differ, a property which is lawlike connected with the first. Since the bias of a coin is one of its properties, it must be connected with some other property such as the position of the coin's centre of gravity or its non-flatness. If no such properties can be detected, it may be presumed that the coin is fully symmetric and unbiassed. It can then be asserted, using the same principle and because the propensity is also a property, 'that there is no asymmetry either in the propensity . . .'. It follows that heads and tails are equally probable in any toss since this is the only symmetric distribution of the chances.

Let us turn at last to our basic postulate in the light of the foregoing discussion. We are concerned with the probabilities of the occurrence of energy eigenstates in a substantially isolated system which has reached the condition of statistical equilibrium. Let the number of eigenstates be denoted $W$. As was said earlier, quantum mechanics shows that to be a finite number. Even so, in the case of any macroscopic quantity of material, it is immensely large. For example, it is of the order of $10^{10^{24}}$ in the case of a gram mole of helium at 273 K and atmospheric pressure.

Now, if the system in question had a *precise* total energy $E$ the value of $W$ would be simply the degeneracy of the eigenvalue $E$. Due, however, to the energy–time uncertainty relation, as well as to the occurrence of external perturbations, the energy of the system lies in a narrow range $E$ to $E + \delta E$. Thus $W$ is the sum of the degeneracies of the eigenvalues in that range. We ask: What is the justification for supposing that all of the $W$ eigenstates are equally probable?

Now, the presence of symmetry within the $W$ possibilities is less obvious than it is in the geometrical example of the six faces of a die. Also one cannot carry out direct frequency tests to check for the absence of bias. On the other hand, no ambiguities about partitioning, or about sub-alternatives, suggest themselves. It is on these latter grounds that the use of the Principle of Indifference seems justified as an heuristic device. Yet it remains heuristic! What can be said is that if the postulate were seriously in error, in regard to significant sub-sets of eigenstates, it seems unlikely that the resultant theory would have been so remarkably successful in all the great variety of its applications. Putting the matter in terms of Mellor's argument, it may be said that none of the properties of substances which are lawlike-connected, through the theory, with the assumed distribution of eigenstates, suggest that this distribution is anything other than symmetric.

The postulate in question thus has the status of a conjectural 'model' about physical reality. In that respect it is like any other hypothesis in science and is partially justified, although not confirmed, by its success.

Of course, this justification is *a posteriori* and it is conceivable that some alternative postulate would be equally successful. For this reason it would be of great value if there were some *a priori* demonstration of the reasonableness of the postulate within the axiomatic system of quantum mechanics. Some authors have in fact attempted to provide this by using the principle of microscopic reversibility in conjunction with the ergodic theorem (Mayer and Mayer, 1940; R. T. Cox, 1955). Yet another demonstration is put forward in Appendix 2.2 where the deterministic temporal development which would be expected from the Schrödinger equation is avoided through the effects of perturbations originating from outside the system. Under these circumstances of slightly imperfect isolation, and if the perturbations occur at random temporal intervals, it is found that all eigenstates within the allowed range are equi-probable.

Before ending this section, it will be useful to clarify the notion of *accessibility* and also to say a little more about the *ensembles* used by Gibbs.

The accessible quantum states are those which the given system, specified by its macroscopic variables, can attain during the period of time for which the system is under study. By 'attain' one means consistently with the axioms of quantum mechanics after allowing for the influence of the externally induced perturbations. Of course, the period must be long enough for a very large number of molecular collisions or transitions to have occurred, since otherwise a representative number of quantum states would not be 'sampled' during a measurement. But the very fact that measurements of macroscopic properties at equilibrium are experimentally reproducible is a clear indication that a fair sample is normally achieved.[12]

Yet the period of time during which certain quantum states are accessible is not necessarily long enough for the attainment of complete thermo-dynamic equilibrium. A familiar example is that of a mixture of hydrogen and oxygen at room temperature and in the absence of a catalyst. The quantum states of water molecules are not regarded as accessible, since to attain them would require aeons. By contrast, the translational, rotational and vibrational states of polyatomic gases are fully accessible to each other – although not *all* such states may be accessible due to certain selection rules. To take yet another example, we may have a gaseous system divided into two parts by means of a partition. Since permeability is a relative term, if we idealise the system by regarding the partition as being entirely impermeable, we shall be taking all those quantum states as being inaccessible which could, in fact, be realised if the partition were slightly permeable during a sufficient duration.

Accessibility is thus a matter of the rates of processes relative to a period of interest. For purposes of simplicity, real systems have to be idealised by supposing that quantum states divide into two sharply distinct classes: (a) those which may be taken as entirely inaccessible during the given period; (b) those which are so readily accessible that a representative sample of them are passed through during that time. The idealisation thus amounts to assuming very rapid rates of transition between certain states, and zero rates between all others.

In any application of the theory, the scientist must therefore make a decision about the reliability of this division into classes. A serious error in regard to, say, a time-dependent selection rule or the stability of some molecular species, would be to create a corresponding error in regard to the states to which the equi-probability postulate truly applies. Yet this is a matter of the adequacy of an approximation and is a common enough situation in science. It does not imply the presence of any subjective factor peculiar to statistical mechanics.

As was said earlier, the notion of the *ensemble* was a conceptual device which allowed the use of the frequency interpretation of probabilities. In the case of a substantially isolated system, the appropriate equilibrium ensemble was called *microcanonical* and was a very large collection of imaginary replicas of the real system of interest, all of them isolated and in the same macroscopic state and having the same Hamiltonian, but distributed over their accessible microstates in accordance with the basic postulate. It is to this ensemble, and only to this ensemble, that the postulate applies and it expresses the *microcanonical distribution* of probabilities.

Another kind of ensemble to which attention will be given shortly is the *canonical ensemble*. Here, the real system of interest together with its imaginary replicas are considered as being in thermal equilibrium with the same very large heat reservoir, and therefore with each other. The probabilities of the energy eigenstates can then no longer be taken as equal, but instead they occur over a range – the *canonical distribution* – due to fluctuating energy exchanges between the system and the reservoir. The exponential character of this distribution will shortly be derived from the microcanonical distribution, and thus from the postulate.

If one adopts the propensity interpretation, the ensemble picture is no longer needed. For the propensities then refer to single events or states-of-affairs. And clearly if statistical mechanics is to be of practical use it must be capable of being applied to a single real system! The ensemble is an artifice and, indeed, Gibbs himself (Dover edn, p. 17) accepted it as such. There is no necessity, he said, for an explicit reference to an ensemble; it serves simply 'to give precision to notions of probability'. Clearly he had in mind the frequency interpretation. Using propensities instead, the microcanonical and canonical distributions can be regarded as pertaining to the accessible quantum states of a single system, rather than being spread, as it were, over imaginary copies. With this usage the term 'ensemble' means nothing more than the equilibrium probability distribution.

And, of course, the notion of equilibrium does not refer to the system being 'in' any particular quantum state; on the contrary, at the system's equilibrium condition the system may occupy momentarily *any* of its accessible states. This naturally allows for the possibility of a momentary 'return' to the initial quantum state, although with minute probability.

## 2.4. Outline of the results

The number $W$ of accessible eigenstates can be used to define an entropy analogue appropriate to the microcanonical distribution. This is

$$S_{BP} = k \ln W \qquad (2.14)$$

where the subscripts stand for Boltzmann and Planck who first used this analogue. We shall not seek to justify (2.14) except to remark that $S_{BP}$ is insensitive to the value of $\delta E$ (p. 30) (Griffiths, 1965; Mayer and Mayer, 1977).

Attention will be given instead to an alternative analogue:

$$S_G = -k \sum_j p_j \ln p_j \qquad (2.15)$$

which is appropriate to the canonical distribution. This is the quantal form of Gibbs' preferred analogue, and the summation is over the energy eigenstates accessible to a system when it is in thermal equilibrium with a large heat reservoir.

In passing, it may be noted that if (2.15) were applied to the micro-canonical distribution, where all of the $p_j$ are equal to $1/W$, it would reduce to (2.14). Under other circumstances $S_{BP}$ and $S_G$ differ in their values only very slightly due to the extreme sharpness of the canonical distribution in the case of macroscopic systems.[13] Notice, too, that $S_G$ is zero in the 'pure' case, since just one of the $p_j$ is then equal to unity and the rest are zero.

Now, in order to make use of (2.15) the $p_j$ have to be evaluated. By regarding the system and the heat reservoir together as a single isolated system, the basic postulate can again be applied and it can thus be demonstrated (as is done in Appendix 2.3) that at equilibrium the $p_j$ are related to the energy eigenvalues $\varepsilon_j$ by the equation[14]

$$p_j \propto \exp(-\beta \varepsilon_j)$$

It is also shown in the appendix that the coefficient $\beta$ has the same value for all systems which are in thermal equilibrium with the reservoir and with each other. Since temperature is the only property that two or more systems in thermal equilibrium necessarily have in common, $\beta$ must evidently be a function of temperature only. This point will be returned to.

Because $\sum_j p_j = 1$, the proportionality constant in the preceding expression can be readily obtained. The results are:

$$p_j = Q^{-1} \exp(-\beta \varepsilon_j) \qquad (2.16)$$

where

$$Q = \sum_j \exp(-\beta \varepsilon_j) \qquad (2.17)$$

$Q$ is the *partition function* and is a sum over all energy eigenstates of the system in question, including degenerate states. Alternatively $Q$ can be taken as a sum over energy levels if each term exp $(-\beta\varepsilon_j)$ is multiplied by $d_j$, the degeneracy of the eigenvalue $\varepsilon_j$.

Equation (2.16) gives the quantal form of the *canonical distribution* of Gibbs. Its great value lies in its exponential structure. This reconciles the multiplicative properties of probabilities $p_i$ and $p_j$, if these are independent, with the additive properties of separable energies. Thus $p_i p_j \propto$ exp$-\beta(\varepsilon_i+\varepsilon_j)$. Equation (2.16) also gives the probabilities as *continuous* functions of the $\varepsilon_j$ (even though the latter may vary discretely), whereas in the microcanonical distribution the $p_j$ are discontinuous since they fall abruptly to zero outside the specified energy range $E$ to $E+\delta E$. Another advantage of the canonical distribution is that the partition function is often more readily calculable than is $W$, the number of accessible energy eigenstates. (Indeed, the value of $W$ already quoted for helium was calculated from the value of $Q$.) And, again, the 'system in a heat bath' corresponds to the thermodynamic variables $T$, $V$, $N_i$ (where $N_i$ is the number of molecules of the $i$th species) and these are more useful in experimental situations than are the variables $E$, $V$, $N_i$ which specify an isolated system.

Within our programme of inquiry we need to demonstrate that no subjective factors are involved in showing that $S_G$ is a satisfactory entropy analogue for equilibrium states. For this purpose let $U$ be the expectation value of the internal energy of the system in question. Its value is given by (2.13):

$$U = \sum_j p_j \varepsilon_j$$

where the $p_j$ refer to the probabilities of energy eigenstates whose eigenvalues are $\varepsilon_j$. The summation is over the states $j$. If this equation is differentiated we obtain

$$dU = \sum_j p_j \, d\varepsilon_j + \sum_j \varepsilon_j \, dp_j \qquad (2.18)$$

Now the two groups of terms on the right-hand side can be identified as a reversible work effect and as a heat effect respectively. This requires the use of the so-called adiabatic theorem[15] in quantum mechanics which asserts that the probabilities of the energy eigenstates remain constant when changes of the externally imposed variables of the system, such as its volume, are made slowly enough. This, of course, is one of the conditions for reversibility in thermodynamics.

The adiabatic theorem is derived in Appendix 2.4 and will here be taken for granted. What it amounts to is that a heat intake, $dq$, by a system is interpreted as being a changed population of the various energy eigenvalues which remain constant. Thus

$$dq = \sum_j \varepsilon_j \, dp_j \qquad (2.19)$$

A reversible performance of work on the system, on the other hand, is identified with a change, $\sum_j p_j \, d\varepsilon_j$, involving the energy eigenvalues themselves at fixed populations.

By rearranging (2.16) we have:

$$\varepsilon_j = \frac{-1}{\beta} (\ln p_j + \ln Q)$$

and if this is substituted in (2.19), recalling that $\sum_j dp_j = 0$,

$$\beta \, dq = -d \sum_j p_j \ln p_j \qquad (2.20)$$

Now the right-hand side is an exact differential, as will be seen in Appendix 2.4. It follows that $\beta$ is an integrating factor for $dq$. Since in thermodynamics the absolute temperature, $T$, is *defined* as being proportional to the reciprocal of this integrating factor, we can make the identification

$$\beta k = 1/T \qquad (2.21)$$

where $k$ is a constant such that at the ice-point of water $T = 273.15$ K. This important result can be further substantiated by evaluating the partition function for a dilute monatomic gas, as is done in Chapter 4, whereby the constant $k$ (Boltzmann's constant) is shown to be the gas constant per molecule.

Finally, the entropy analogue (2.15) can be shown, using (2.20) and (2.21), to lead to a recovery of the familiar relation by which entropy was first defined in thermodynamics:[16]

$$dS_G = (dq/T)_{rev.} \qquad (2.22)$$

It may be added that Gibbs showed that $S_G$ has a maximum value for the canonical distribution (2.16); any other distribution giving the same mean energy would result in a lower value of $S_G$. It was thus implicit in Gibbs that the analogue $S_G$, unlike the thermodynamic entropy, should be applicable to non-equilibrium situations. As Kac (1959) put it, the Gibbsian notion of equilibrium does not refer to the *state* of a body but rather to a particular probability distribution – that which maximises $S_G$.

For later use it will be helpful to have expressions for the internal energy and entropy in terms of the partition function. These are readily obtained from the foregoing equations:

$$U = kT^2(\partial \ln Q/\partial T) \tag{2.23}$$

$$S = k \ln Q + kT(\partial \ln Q/\partial T) \tag{2.24}$$

where the partial differential coefficients are at constant volume and constant mole numbers. The corresponding formula for the Helmholtz free energy is

$$\mathscr{A} = -kT \ln Q \tag{2.25}$$

That this is the simplest of the expressions is, of course, no accident, since $\mathscr{A}$ is the 'characteristic function' for a system having prescribed volume and temperature (rather than energy) and these are the same conditions as apply to the canonical distribution. All other thermodynamic relations, such as those concerning the heat capacities, may be readily obtained, in terms of the partition function, from (2.23)–(2.25).

In summary, it has been shown that the change of the entropy analogue $S_G$ can be equated, in the case of reversible processes, to the change of the thermodynamic entropy without any appeal to 'ignorance' or to 'lack of information'. However, it remains to be seen whether $S_G$ has the increasing properties of thermodynamic entropy and this question will be deferred to Chapter 3.

Before ending this section, it is of interest, from a philosophical viewpoint, to retrace our path from §2.2 onwards, particularly in regard to the successive coming and going of various 'theoretical terms'. The matter started with the state vector and was followed by operators and by energy eigenstates and eigenvalues and the quantum probabilities. Since the systems to which statistical mechanics is intended to apply are usually specified only in regard to their macroscopic variables, it is impossible to use the theoretical apparatus of quantum mechanics for an actual calculation of the probabilities. For this reason it was necessary to adopt an additional postulate limited to equilibrium conditions. This allowed the probabilities to disappear from the scene and resulted in the partition function which is the culmination of the Gibbsian form of statistical mechanics. Out of all the original theoretical terms the partition function contains only the energy eigenvalues and their degeneracies, and these, too, can be eliminated in especially simple situations where the partition function can be calculated by use of the quantum mechanical axioms. More usually one has to make use of experimental data, such as is obtainable from molecular spectra, or to adopt various approximations as in the treatment of the solid state. The significant point is that it was the original theoretical terms which legitimised the foundations of the quantum statistical mechanics. This gives results which are in better agreement with experiment than if the statistical theory had been based on classical mechanics.

## 2.5. Entropy, 'disorder' and 'ignorance'

One of the reasons why entropy has been regarded as subjective lies in its alleged connection with 'disorder'. Let us return to the passage from Maxwell quoted in §1.1:

> Dissipated energy is energy which we cannot lay hold of and direct at pleasure, such as the energy of the confused agitation we call heat. Now confusion, like its correlative term order, is not a property of things in themselves, but only in relation to the mind which perceives them.

The final sentence of this quotation can be readily accepted. The 'orderliness' of things depends on the possibility of our seeing in them some sort of pattern. An 'ordered' pack of cards is perceived to have such a pattern, although the particular sequence which is regarded as orderly may well depend on the rules of some particular card game. From an objective standpoint the sequence having an apparent orderliness is just one out of 52! possible sequences, as numbered from top to bottom of the pack. 'Orderliness' is both relative and subjective.

We can also agree with Maxwell that no 'orderliness' is apparent in the thermal agitation of molecules. However, there are other examples in which the supposed connection of entropy with disorder is very dubious indeed. Some authors have declared that when a gas expands from one half of a container into the whole of that container there is an increase of *disorder* due to the molecules becoming more widely dispersed. But it can equally well be argued that there is an increase of *order* in this process because the system becomes more uniform. Whereas in the initial state the gas is *un*symmetrically distributed between the two halves of the container, in the final state the gas is very nearly symmetrically distributed – indeed, exactly so as far as human powers of discerning 'pattern' are concerned. Thus entropy increase, on this argument and in this particular example, is accompanied by increase of *order*. Similar remarks can be made about spontaneous processes in which heat, electricity, etc., become more symmetrically distributed between two or more bodies.

Another counter-example to the prevailing mythology is provided by the spontaneous crystallisation of a super-cooled melt. If this process is allowed to occur under adiabatic conditions the entropy of the system increases. But can it reasonably be claimed that the crystalline lattice is a more disorderly structure than was the super-cooled melt? Surely not. In this important instance the only way in which the myth can be 'saved' is by supposing that 'orderliness' can be separated into two parts, configurational and thermal. The gain of order which accompanies the formation of the lattice is more than offset, in this view of the matter, by an increase of disorder due to a randomisation of the liberated potential energy (i.e. the latent heat) over the vibrational modes of the crystal. In other words, 'disorder' must refer to the distribution of energy over energy levels as well as to the distribution of particles over positions in space. Yet this is a highly sophisticated way of looking at the matter and is theory-dependent. It is remote from any simple human perception of a loss of pattern.

The inference to be drawn from these examples is that entropy cannot be interpreted in a reliable and comprehensive manner as increase of 'disorder'. Indeed, the latter notion is almost useless outside the highly subjective situations where some particular group of humans recognises the presence or absence of a significant pattern in the things they are perceiving.[17]

If a verbal interpretation of entropy is required, a far more reliable one is to be found in the notion of 'spread' as used by Guggenheim (1949). An increase of entropy may be said to correspond to a 'spreading' of the system over a larger number, $W$, of occupied quantum states. Alternatively one might say that entropy is a measure of the extent to which the system in question is unconstrained; the less constrained it is, the greater is the number of its accessible quantum states for given values of those constraints which exist. This view of the matter is not at all subjective, since the quantum states do not exist merely 'in relation to the mind which perceives them' (to use Maxwell's phrase) but are generally assumed to correspond to fully objective states of affairs.

Let us turn to the supposed connection between entropy and 'ignorance' or 'lack of information', a matter which will be taken up again in Chapters 3 and 5.

Max Born, it will be recalled, said that 'Irreversibility is a consequence of the explicit introduction of ignorance into the fundamental laws.' Although irreversibility has not yet been discussed it can be seen already that Born's attitude to statistical mechanics is seriously misleading. As was emphasised earlier, the objective of statistical mechanics is to support, and to go beyond, thermodynamics; for this reason statistical mechanics is concerned with systems which are no more than *macroscopically* specified. These are not necessarily large systems (although they often are); they are rather those systems whose precise quantum state is not available. This is so not because of an 'explicit introduction of ignorance', but only because of the need for *a body of theory which will apply to the results of the simplest kinds of measurements*, e.g. of temperature, volume and composition. As was said in §2.3, this is a more correct viewpoint than that of regarding statistical mechanics as bringing 'ignorance' into the fundamental laws or of not using information which might have been obtained by going beyond those simple measurements.

A further important point is that irreversible processes, such as temperature equalisation or the mixing of gases, take place in the same way, however closely we, as their observers, *might* try to specify them. The temporal asymmetry of macroscopic phenomena (on which the entropy concept is founded) has nothing to do with *our* amount of information.

The unwarranted intrusion of a subjective attitude is nicely illustrated by a passage from von Neumann (1955, p. 400). Entropy increase, he said, is due to the fact that the observer cannot measure 'everything which is measurable in principle. His senses allow him to perceive only the so-called macroscopic quantities.' He further maintained that 'for a classical observer, who knows all coordinates and momenta', the entropy is zero.

The 'entropy' he speaks of cannot be the thermodynamic entropy, or the normal statistical mechanical entropy. For these, as has been said, are *intended* to apply to systems specified only in terms of *the macro-variables*. If we also knew 'all coordinates and momenta' this additional information would not change the values of the thermodynamic and statistical mechanical entropies in the least. In particular it would not make those values equal to *zero* as von Neumann said. For consider some body at a finite temperature; however well we knew its atomic state, its entropy could always be made smaller, but without necessarily becoming zero, by reducing its temperature.

Or again, if we consider the matter from the standpoint of the Boltzmann–Planck analogue, $W$ is the number of accessible quantum states for *the specified energy, volume and composition*. This is what the theory is designed to give and *not* the number 1 which would be obtained from the complete atomic specification envisaged by von Neumann. The latter is quite irrelevant to the macroscopic description.

What would be appropriate is a statement quite different from von Neumann's. If the temperature of a gas or liquid is lowered sufficiently it may well become a crystalline solid; as the absolute zero is approached the occupied quantum states of the crystal crowd into the ground states and thus the quantum state of the crystal as a whole may be said to become more readily 'knowable'. Thus, inverting von Neumann's statement, it becomes more feasible for 'a classical observer' to determine the coordinates and momenta.

The 'ignorance' view of entropy has also been strongly expressed by Jaynes (1965). He remarked, as was mentioned earlier, that entropy can be said in a subjective sense to measure 'our *degree of ignorance* as to the true unknown microstate, . . .'.

This is correct but the effect of the remark is to draw attention to an irrelevance. It is true that we do not know which particular one, out of all the $W$ accessible quantum states of an isolated system, may be said to be occupied at any instant. Yet this has nothing to do with the objective physics of the situation! For the actual situation is one in which the system is specified *only* in regard to having a given volume and composition, and internal energy lying between $U$ and $U + \delta U$. That being so, the occurrence of the number $W$ arises quite naturally out of the statistical mechanical theory, and without reference to human powers of knowing. It is thus a quite unnecessary gloss on the significance of the analogue $S_{BP} = k \ln W$ to make it seem *dependent* on those powers.

Let it be remembered that $W$, in quantum terms, is essentially a *degeneracy* – it is the number of accessible and independent energy eigenstates of the given system.[18] Consider a single molecule and suppose that the degeneracy of one of its eigenvalues has the value 2. The fact that it is unpredictable which of the two eigenstates will be observed as the result of a quantum measurement is not normally regarded by scientists as being a subjective matter. There is no good reason why we should take a different attitude to $W$ which is the degeneracy of a system composed of a multitude of molecules.

In short, what we might be able to know by means of quantal measurements on micro-observables is irrelevant to the kind of theory, namely statistical mechanics, which is specially designed to cope with situations where only macro-observables are used. And in regard to irreversibility the subjectivist is driven to the absurd conclusion, as Popper (1982) put it, that molecules escape from a bottle 'because we do not know all about them, and because our ignorance is bound to increase unless our knowledge was perfect to begin with'.

To be sure, the notion of 'information' can be used in a highly technical and non-subjective sense. The extent to which 'information', when so regarded, can be taken as an inverse measure of entropy will be discussed in Chapter 5. For the present it is sufficient to notice that a set of instruments and a pre-programmed computer could evaluate the entropy of a substance and that this could be achieved, conceptually at least, in man's absence from the world. The data alone is sufficient.

Yet there is a further version of the 'ignorance' view of entropy; namely that our knowledge of the world is inevitably 'coarse grained' and macroscopic; if we were actually as small as the atoms and molecules, it has been said, we would be quite unaware of the occurrence of any irreversible processes.

This is surely just as mistaken a view as others which have been discussed. Suppose that we were indeed as small as the atoms and molecules whilst still possessing certain faculties – in particular the faculty of being able to distinguish between nearest neighbour molecules of different species. We could then observe, at the molecular level, the mixing of a gas of $A$-type molecules with a gas of $B$-type molecules, and we could also discover that the reverse process of 'unmixing' seldom occurs. In short, macroscopic processes have microscopic counterparts, and 'micro-beings' could therefore know about irreversibility. But, of course, this is all very hypothetical and more will be said in Chapter 5 about the limitations on what could be achieved by a 'demon' operating at the micro-level.

Finally, let us recall from §1.3 that there is a very satisfactory convergence between statistical mechanical entropies obtained from spectroscopic data, on the one hand, and Third Law calorimetric entropies on the other. This could only be made consistent with the subjective view of entropy by claiming, contrary to all the evidence, that heat and temperature are also subjective, since it is these which determine the values of the calorimetric entropies. It would be hard to explain the remarkable degree of agreement provided by two such radically different methods unless they correspond to objective reality.

## 2.6. Is entropy a property?

We have thus concluded that entropy has a fully objective status. It is a function of the variables which are used in the thermodynamic type of specification and is just as objective as is, say, a refractive index, a density or a vapour pressure which are similarly specified.

Yet this very comparison raises the question whether entropy can be regarded correctly as being a *property* of a body. This issue can best be discussed by adopting the purely thermodynamic viewpoint.

Now, in thermodynamics one can only obtain the entropy differences, $S_2 - S_1$, between states 1 and 2 of a body. Whereas normal macroscopic properties, such as refractive indices, have 'possessed' values in each of the states 1 and 2, this is not the case with entropy – or at least it is not the case unless one supposes that entropy has a natural zero at the absolute zero of temperature. Yet this particular reading of the Third Law does not provide a very secure base line if only because there is a current lack of knowledge about entropy within the nucleus. Only if it turns out that nuclear entropy, if it exists, diminishes to zero at zero temperature, might we be justified in assigning absolute or 'possessed' entropy values to bodies at any higher temperatures.

The peculiarity of entropy is brought out even more clearly by the fact that the difference $S_2 - S_1$ can be measured only along a reversible route between states 1 and 2. It is indeed a very unusual sort of property whose value can only be determined as a difference along a special kind of path!

Yet it would be contrary to normal scientific usage to say that the entropy of a substance is *not* one of its properties. The matter can be cleared up by noticing the special class of properties known as *dispositional*. A familiar example already quoted in § 2.3 is the fragility of glass; this property of glass is a disposition to break under certain circumstances, but otherwise is not an 'observable'.

Similarly, the difference $S_2 - S_1$ is a measure of the disposition of a body to give rise to an externally measurable heat effect when, but only when, the integral $\int_1^2 dq/T$ is taken along a reversible path. If the change is not reversible the difference $S_2 - S_1$ (although remaining the same for the body itself between the fixed states 1 and 2) does not make itself manifest as the same intake of heat. Indeed, in the case of a spontaneous change $1 \rightarrow 2$ within an adiabatically isolated system, there is no external manifestation of heat. As will be seen in Chapter 4, this point about entropy failing to manifest itself as $\int dq/T$ when the path is irreversible plays an important part in the resolution of Gibbs' Paradox.

# 3

# *Coarse graining*

### 3.1. Introduction

Although only equilibrium states have been discussed so far, it might be hoped that statistical mechanics would have something useful to say about non-equilibrium situations as well; in particular that it would give answers to a number of questions, all closely related. For example: Why do systems mostly tend towards an equilibrium state? Why are natural processes time-asymmetric and irreversible? How is entropy increase to be accounted for?

Since the scope of this book is limited to the subjectivity/objectivity issue, no attempt will be made to survey the whole vast field of non-equilibrium statistical mechanics. Neither shall we deal with Boltzmann's original $H$-theorem, since this was limited to the case of dilute gases. Attention will only be given to the mathematical device known as coarse graining, as used in the proof of the generalised $H$-theorem, because this device has often been held to involve subjective factors. (See, for example, Tolman, 1938, p. 561, and references in §3.4.)

The familiar difficulty which statistical mechanics has to confront when seeking to demonstrate irreversibility is that the basic laws and theories of physics are all $t$-invariant. This means that elementary processes such as collisions, quantum transitions, etc., have precisely the same frequency of occurrence whether they occur from a micro-state 1 to a micro-state 2 or from the 'time-inverted' state of 2 to the time-inverted state of 1 in the same temporal interval.[19] How, then, can it occur that a macroscopic process, involving immense numbers of elementary processes, can show temporal asymmetry? This was, of course, the basis of the famous 'objections' of Loschmidt and Zermelo to Boltzmann's theorem.

Before coming to coarse graining let us consider these 'objections' in the classical context in which they were originally put forward. That is to say, in the context in which the precise coordinates and momenta of particles are supposed to be knowable, at least in principle. Consider a gas and let $\{A_i\}$ and $\{B_j\}$ be those sets of exactly specified microstates which are accessible to the gas in its initial and final thermodynamic states, $A$ and $B$, respectively. For example $\{A_i\}$ might refer to the set of states before the gas expands into a vacuum and $\{B_j\}$ might refer to the set of states after expansion is complete. Let $\bar{A}_i$ and $\bar{B}_j$ differ from $A_i$ and $B_j$ respectively only in that each molecule has an exactly reversed velocity vector. Since states $A_i$ and $\bar{A}_i$ have the same energy it may be assumed that they are equally likely to occur and similarly for states $B_j$ and $\bar{B}_j$. Thus it would appear, in accordance with the *t*-invariance of the basic laws, that, if the *macroscopic* process $A \rightarrow B$ occurs, the reverse process $B \rightarrow A$ should occur equally frequently in the same direction of time. On these grounds Loschmidt argued that Boltzmann's *H*-theorem must be in error.

Yet this argument does not do justice to the *statistics* of systems containing a very large number of molecules. For out of the set $\{\bar{B}_j\}$ of inverted velocities only a very small fraction will be suitable for allowing a gas which has expanded into a larger volume to subsequently contract momentarily into its original smaller volume. In short, the existence of a set of suitably oriented velocities which would allow of recontraction becomes exceedingly improbable when the system contains a large number of molecules.

It was on these grounds that Boltzmann replied to Loschmidt.[20] More detailed treatments have been given by Balescu (1967) and by Prigogine *et al.* (1973). It may be noted, however, that something similar to what Loschmidt had in mind can actually be observed in the highly exceptional circumstances of Hahn's spin echo experiment which is described in Appendix 3.1. The situation here is that the set of the type $\{\bar{B}_j\}$ contains the same number of members as the set of the type $\{A_i\}$; for every original spin there is a spin with a reversed velocity of precession. Furthermore, all of these velocities can actually be reversed simultaneously by applying a magnetic pulse. As a consequence an almost complete 'return' to the initial state can, in fact, be observed. Yet the circumstances are so abnormal as to be of little significance in the present context.

Unlike Loschmidt's 'objection', the one due to Zermelo is much more far-reaching and absolute in character. But it has to be seen in an appropriate temporal perspective.

If one were to release just two molecules, say, into an empty tube they would quickly find their way into all parts of the total volume. Clearly they will eventually be momentarily present once again close to the point of inlet. So, too, with 3, 4, ..., molecules but the 'recurrence time' for their simultaneous return to the inlet point may be expected to increase very rapidly with increase in the number of molecules involved. This is the essence of Poincaré's recurrence theorem in classical mechanics; any finite isolated system must eventually return to a state arbitrarily close to its initial state. A similar theorem holds in quantum mechanics and in this case the statistical predictions are also recurrent, since the states are discrete. (Ono, 1949; Percival, 1961, 1962; Hobson, 1971.)

Boltzmann fully accepted Zermelo's argument that, because of the Poincaré recurrence, there can be no absolute irreversibility in finite systems which are perfectly isolated. But he pointed out that aeons of time would be required for recurrence to obtain in macroscopic systems. 'It will suffice to have a little patience', Poincaré is reported to have said, and perhaps he meant it flippantly. For it can be readily calculated from Einstein's fluctuation theory that the entropy of, say, one gram mole of helium can be expected to fall by a mere millionth below its equilibrium value just once in $10^{10^{19}}$ years (Denbigh, 1981b, p. 106). Complete recurrence in Poincaré's sense would require a vastly longer period. It follows that effective irreversibility can be confidently expected in isolated macroscopic systems during all periods of time which it makes sense to consider. In such systems the statistics of large numbers of particles far outweighs the *t*-invariance prevailing at the atomic level.[21]

Even so, a rigorous proof that entropy does indeed tend to increase during all 'reasonable' durations – that is to say, a proof of the Second Law at a level deeper than is obtainable from straight thermodynamics – has been found to be one of exceptional difficulty. The basic reason is a restriction imposed by Liouville's theorem which arises from the *t*-invariant equations of motion in conjunction with a conservation law. It was to avoid this restriction that the notion of coarse graining was adopted.

### 3.2. Gibbs' treatment of irreversibility

Coarse graining can best be approached by recalling how Gibbs (1902) tried to solve the problem of irreversibility within the context of classical statistical mechanics. It may be added that the problem is not significantly different in quantum statistical mechanics since here, too, there is a restriction due to the quantal analogue of the Liouville theorem.

Consider a system of individual elements, such as molecules, each having *f* degrees of freedom. The state of each element may be described classically by specifying its *f* coordinates and its *f* momenta, and this description is called a description in $\mu$-space. If the whole system contains *N* such elements its description requires a specification of *Nf* coordinates and *Nf* momenta, and this is said to be a description in $\Gamma$-space. The state of the whole system may thus be represented by a single point, *P*, in the 2*Nf* dimensional $\Gamma$-space. The point moves about erratically, even when the system is in equilibrium, due to the effect of small but variable external influences. As already discussed, these influences will disturb the state of the system, even though it may be substantially isolated.

We here adopt Gibbs' concept of the ensemble, i.e. an imaginary collection of $\mathcal{N}$ such systems, replicas of the single real system of interest except in so far as they differ from each other in regard to their momentary coordinates and momenta. They all have the same energy but momentarily occupy different points $P_\alpha, P_\beta, \ldots, P_\mathcal{N}$ in the $\Gamma$-space. Let $\delta\Gamma$ be a $2Nf$ dimensional element in this space:

$$\delta\Gamma = \delta q_1 \ldots \delta q_{Nf}\, \delta p_1 \ldots \delta p_{Nf}$$

Let $\rho$ be a density function, $\rho(P, t)$, such that $\mathcal{N}\rho\delta\Gamma$ is the number, $\delta\mathcal{N}$, of systems in the volume element $\delta\Gamma$ centred on some point $P$ at time $t$, and such that the integral of $\mathcal{N}\rho\delta\Gamma$ over the whole space is equal to $\mathcal{N}$. The density $\rho(P, t)$ may be regarded as the probability per unit volume of the $\Gamma$-space that a system picked at random from the ensemble will be in the particular region $\delta\Gamma$.

Suppose that the processes occurring in the system of interest are such that the number of degrees of freedom remains unchanged.[22] Using the Hamiltonian equations of motion it can then be proved that $\rho$ remains constant in the close vicinity of any one of the systems of the ensemble as this system moves within the $\Gamma$-space. This is Liouville's theorem and it can be expressed as

$$d\rho(P, t)/dt = 0 \tag{3.1}$$

This total derivative refers, of course, to 'following the flow' in the $\Gamma$-space. By contrast, it is the partial derivative $\partial\rho/\partial t$ which refers to a *fixed position* in the space and $\partial\rho/\partial t = 0$ only obtains when there is statistical equilibrium.

Gibbs proceeded (using a different notation) to investigate the properties of the function:

$$S'_G = -k \int \rho \ln \rho \, d\Gamma \tag{3.2}$$

where $S'_G$ is the classical equivalent of the quantal $S_G$ defined in equation (2.15) and the integration is over the whole accessible region of $\Gamma$-space. By using the classical equivalent of the canonical distribution, (2.16), Gibbs obtained, in a non-quantal form, all of the results for equilibrium states which have already been described in §2.4.

But what about the non-equilibrium situations, the irreversible processes, which are our present concern? Gibbs fully understood the problem created by Liouville's theorem. Since (3.1) holds for every representative point $P$, the integral in (3.2) remains constant in time. It follows that the entropy $S'_G$ *cannot increase*, consistently with the theorem, during the gradual unfolding of an irreversible process.[23] As he put it, 'we find by this method no approach toward statistical equilibrium in the course of time.' (*Loc. cit.*, p. 144.)

Gibbs put forward a tentative alternative method based on his well-known analogy of the spreading of a quantity of colour during the stirring of a liquid.[24] He pointed to an apparent paradox: although Liouville's theorem shows that 'the density of the coloring matter at any same point of the liquid will be unchanged, . . . yet no fact is more familiar to us than that stirring tends to bring a liquid to a state of uniform mixture, . . . .'

This he explained as being a matter of *our* method of evaluating the density of colour. If we were to look at the system on the very fine scale of the colour filaments we would see no mixing, but if we were to regard it on a much coarser scale an approach to uniformity would be apparent. Here there was clearly the beginning of a distinctively *subjective* approach to the understanding of irreversibility. Gibbs had put forward the notion of *coarse graining* as a device for achieving this understanding and this was subsequently developed by the Ehrenfests (1912) in classical statistical mechanics and by Tolman (1938) and others in quantum statistical mechanics.

### 3.3. The generalised *H*-theorem

In the classical treatment of this theorem by the method of coarse graining, the $2Nf$ dimensional $\Gamma$-space was divided into regions each of volume $\tau$.[25] Let us call them 'stars' (by analogy with the 'Z-stars' of the Ehrenfests) in order to distinguish them from the 'cells' in $\mu$-space as used in Boltzmann's original *H*-theorem. Similarly in the quantal treatment (Tolman, 1938), one considers *groups* of energy eigenstates and not the individual eigenstates.

In justification of coarse graining, it has been said that the position of a system in classical $\Gamma$-space cannot be determined by experiment with unlimited accuracy. And neither can its precise eigenstate be experimentally established. The volume $\tau$ of a star, or the number of eigenstates in a group, should therefore (it is claimed) be such as corresponds to 'the limits of accuracy actually available to us' (Tolman, p. 167).

This point – the fact that there are no laws of physics which determine the size of the stars or quantum groups – is one of the sources of the view that coarse graining is subjective (cf. van Kampen, 1962, and Hoyningen-

Huene, 1976). Closely related is the supposed 'loss of information' as the systems of the ensemble redistribute themselves among the stars during the course of an irreversible process. (Of course, coarse graining is not needed in the treatment of equilibrium, since one can here adopt a stationary microcanonical or canonical distribution of probabilities.)

Let us leave this aside for the moment and proceed with the generalised $H$-theorem. Here one defines a *coarse grained density*:

$$\bar{\rho}(\tau, P, t) = \frac{1}{\tau} \int_{\tau} \rho(P, t) \, d\Gamma \qquad (3.3)$$

where the integral is over the volume $\tau$ of a star and $\rho$ has the same meaning as in (3.1) and (3.2). The coarse grained density in a star is thus the mean of the fine grained densities in the vicinity of the point $P$ on which the star is centred. A quantity $\bar{H}$ is also defined:

$$\bar{H} = \int \bar{\rho} \ln \bar{\rho} \, d\Gamma \qquad (3.4)$$

where the integration is now over the whole accessible region of phase space.[26]

$\bar{\rho}$, unlike $\rho$, is not subject to the restriction expressed by equation (3.1). It follows that $\bar{H}$ need not remain constant in time in the way that Liouville's theorem requires $S'_G$ to remain constant. Let it be supposed that the system of interest is at equilibrium at some initial instant $t_0$ and that a measurement of sufficient accuracy is made to locate that system in a particular star. All the imaginary replicas of the ensemble must also be located in that star. Furthermore, the postulate of equal *a priori* probabilities allows it to be asserted that $\rho_0$ is constant throughout the volume of the star, although zero elsewhere. Thus

$$\rho_0 = \bar{\rho}_0 \qquad (3.5)$$

Immediately following $t_0$ some constraint on the system is lifted (e.g. a partition is removed) so that an irreversible process is initiated. As a consequence (3.5) does not continue to hold at times later than $t_0$; indeed, the whole point of coarse graining is the supposition that the members of the originally occupied star will enter different stars subsequent to $t_0$ and may become widely separated in the $\Gamma$-space. Their representative points continue to provide a value of the density $\rho$ which is constant 'following the flow', as is asserted by (3.1), and therefore those representative points continue to remain within the same volume $\tau$ in spite of their migrations. This means that the constant volume $\tau$ which contains the original points has greatly changed its 'shape' and has developed 'filaments' as in Gibbs' colour analogy. It follows that at later times, $t_1, t_2$, etc., there will be regions of $\Gamma$-space where $\rho \neq \bar{\rho}$.

It is then easy to prove, as is done in Appendix 3.2, that at the instant $t_1$:

$$\bar{H}_0 \geqslant \bar{H}_1 \qquad (3.6)$$

Similarly at the instant $t_2$:

$$\bar{H}_0 \geqslant \bar{H}_2 \qquad (3.7)$$

and so on. In the case of an isolated system it can also be shown that a coarse grained entropy can be defined as

$$S_{cg} = -k\bar{H} \qquad (3.8)$$

such that $S_{cg}$ is an entropy analogue.

It will be noticed that the foregoing does not prove that $\bar{H}_1 \geqslant \bar{H}_2$. Putting the matter in terms of entropy, the theorem does not demonstrate that entropy will steadily increase. All that can be said is that it seems reasonable to expect *a high probability* for $\bar{H}$ to diminish, and for $S_{cg}$ to increase, due to a continued 'spreading' over the accessible region of $\Gamma$-space until a condition of equilibrium is attained. In the case of an isolated system the equilibrium condition corresponds, of course, to a uniform spread over the constant energy surface in the $\Gamma$-space. In the 'ensemble' picture this is equivalent to the microcanonical distribution of probabilities of §2.3.

The quantal development of the generalised $H$-theorem proceeds in a similar manner using the groupings of eigenstates already referred to, and similar conclusions are reached.[27]

### 3.4. Coarse graining reviewed

In the context of this book two questions suggest themselves: (1) Is coarse graining really subjective? (2) Is coarse graining an *essential* technique for the discussion of irreversibility in statistical mechanics? One might also ask, although somewhat rhetorically, whether a coarse grained entropy *has any actual use*, theoretical or practical, apart from its being a product of the $H$-theorem.

A number of authorities have been inclined to answer the first question in the affirmative on the grounds, as has been said, that the mode of partitioning the phase space into stars is not determined from within physics, and further that the result $\bar{H}_0 \geqslant \bar{H}_1$ is simply a 'loss of information'. As Tolman (p. 172) put it, the change of $H$ 'corresponds to a decrease with time in the definite character of our information as to the condition of the system of interest'.

Grünbaum (1975) has argued against this view. He admits that the charge of subjectivity is made plausible by the fact that the magnitude of an entropy change depends on the human choice of the size of the stars. One can even find types of partitioning which manifest a diminution of entropy with time! But this fact, he argues, relates only to a single system. As soon as we consider a union, $U$, of ensembles it follows that: 'For any and every partition, the common initial entropy of all the systems in $U$ will either have increased *in a vast majority* of the systems by the time $t + \Delta t$, or it will have remained the same.' In other words, it can be shown that, after averaging over all macrostates relative to a partition, however it might be chosen, the entropy either increases or remains constant. Grünbaum therefore concluded that the objective physical significance of a coarse grained entropy is not impugned by the conceptual process of coarse graining and 'cannot be held to be merely expressive of human ignorance of the underlying microprocesses'.

Although this is an important argument (cf. Sklar, 1977), it remains a fact that the actual *numerical value* of a coarse grained entropy is a function of the chosen size, $\tau$, of the stars and is not determined by nature.

The answer to the second question is also very controversial. A number of authors (Jancel, 1969; Farquhar, 1964; Penrose, 1970; Landsberg, 1978) have regarded the concept of coarse graining as being entirely natural and proper, and also as being an essential tool for the understanding of irreversibility. On the other hand, there are several vigorous critics, scientists who regard the whole notion as being highly dubious. For example, Jaynes (1965) remarked that: 'Any really satisfactory demonstration of the second law must . . . be based on a very different approach than coarse graining.' Prigogine (1980) wrote in a similar vein:

> The view that irreversibility is an illusion has been very influential and many scientists have tried to tie this illusion to mathematical procedures, such as coarse graining, that would lead to irreversible processes. . . .
> None of these attempts has led to conclusive results.

Another important critic was Blatt (1959), who pointed out that the conditions of Hahn's spin echo experiment requires the use of a *fine grained* entropy. Indeed, a coarse grained method of approach to this experiment would be very unsatisfactory since it neglects the correlations which explain the 'echo'. This point is discussed more fully in Appendix 3.1. Blatt went on to suggest that a deterministic Hamiltonian treatment of the approach to equilibrium must be abandoned since no real system is ever perfectly isolated but instead is subject to small external influences which are random.

Such a viewpoint had, in fact, been adopted by Bergman and Lebowitz (1955, 1959), who used a fine grained entropy in conjunction with a Liouville equation *modified* by the inclusion of a new stochastic term. This expressed the probability that any representative point $P$ may be suddenly and randomly displaced from its location in the $\Gamma$-space to a changed location due to sudden impulses originating in the world external to the system in question.

This line of approach, which has also been used by Robertson and Huerta (1970), Fano (1983), and others, seems to us to be very important. Indeed, we have used the same considerations ourselves; for example in § 2.3 we pointed out that the supposition that a physico–chemical system undergoes a deterministic development, in accordance with the Schrödinger time-dependent equation, is entirely unrealistic since *all* systems are subject to random disturbances, however small. In Appendix 2.2 we used the same idea for the purpose of validating the assumption of equal *a priori* probabilities in the equilibrium state.

At the same time, it would be counter-intuitive to suppose that the macroscopic aspects of irreversible processes could not occur *without* the external perturbations. To take an example, a gas would surely expand into a vacuum even if the whole system were *perfectly* isolated. The same applies to processes of mixing, chemical reaction, temperature equalisation and so forth. Even though complete isolation may be impossible to attain, it would seem wrong to suppose that macroscopic irreversibility would not occur if it were attained. It is rather at the micro-level, the destruction of correlations, that Blatt's argument, and the papers of Bergman and Lebowitz, make an important contribution. This will be seen more fully in the next section.

### 3.5. An apparent inconsistency in the theoretical system

We return to the question (top of p. 53) of how Liouville's theorem can be reconciled with the *reality* of the increase in the normal (or fine grained) entropy[28] of statistical mechanics. Indeed, the 'problem of irreversibility' is not so much a matter of understanding *why* entropy increases – for this is fairly clear at the intuitive level – as of legitimising our intuition within a theoretical system which is self-consistent.

A simple example of the apparent difficulty is provided by the irreversible process which consists in the expansion of a dilute monatomic gas from an initial volume $V_1$ to a final volume $V_2$ due to the removal of a partition, or the opening of a valve, which leads to an evacuated chamber.

It is a straightforward thermodynamic result that the entropy increase, as between the initial equilibrium condition 1 and the final equilibrium condition 2, is given by

$$S_2 - S_1 = Rn \ln (V_2/V_1)$$

where $R$ is the gas constant and $n$ is the number of moles of gas. Precisely the same result is obtainable from the statistical mechanics of equilibrium states by using the Sackur–Tetrode equation which will be derived as equation (4.5) in the next chapter. This equation involves the fine grained entropy and it immediately gives

$$S_2 - S_1 = kN \ln (V_2/V_1) \tag{3.9}$$

Here $N$ is the number of atoms and $k$ is Boltzmann's constant; since $N = \mathbf{L}n$ and $k = R/\mathbf{L}$, where $\mathbf{L}$ is the Avogadro number, the two equations are the same.

The question arises how the undoubtedly correct equation (3.9) can be reconciled with the apparent constancy of the fine grained entropy as discussed already. This issue can be illustrated by thinking of the process as occurring in the six-dimensional $\mu$-space ('molecule space') instead of in the $\Gamma$-space. In the initial state, the limits of the *positional* coordinates of each atom are determined by the initial volume $V_1$. However, as soon as the partition is removed, the fastest atoms having suitably oriented vectors start moving into the evacuated chamber and are followed by the slower-moving ones. This, of course, is the reason why the spontaneous process of expansion takes place and it has been argued in §3.1 that it is effectively irreversible.

The corresponding events in $\mu$-space must be thought of a little differently since the *momentum* coordinates are also involved. Any point in this 'space' changes its 'position' with time in a manner prescribed by Hamilton's equations of motion. If $\Omega$ is a region in this space, comprising a set of such points, after a time $t$ these points will have changed their positional coordinates and their momenta so that $\Omega$ has become $\Omega(t)$. Hamilton's equations can be used to show that the volume of $\Omega(t)$ does not change, even though its *shape* may become very distorted, and this constancy of the volume of $\Omega$ is an alternative form of Liouville's theorem. For, if there were initially $n$ atoms in the region $\Omega$, after the time $t$ these atoms will be in the region $\Omega(t)$ and, since this region remains of constant volume, the density $\rho$ also remains constant in accordance with our earlier formulation of the theorem.

For purposes of simplicity, consider atoms which are confined to moving along the $OX$ direction, parallel to the length of the vessels, and having momenta, $p_x$, between zero and some arbitrary upper limit $p'_x$. Let $x'$ be the length of the vessel up to the position of the partition. At the instant $t_0$ when the partition is removed, the region of the two-dimensional phase space defined by $p'_x$ and by $x'$ is shown by the rectangle $\Omega(0)$ in Figure 1(a). At a slightly later moment $t_1$ this region has become the parallelogram $\Omega(t_1)$ which is obtained by plotting $p_x$ against $x$. Its shape is determined by the linear relationship which exists between the distance the atoms have progressed towards the right and the value of their momenta in that direction.

At a later instant, $t_2$, some of the fastest-moving atoms have been reflected at the right-hand wall of the vessel and this effect is shown by the small triangle, which corresponds to negative values of $p_x$, which is part of the region $\Omega(t_2)$. At an even later time $t_3$, when multiple reflections have occurred at both ends of the vessel, the shape of $\Omega$ has become as it is shown in Figure 1(b). Evidently the *shape* in $\mu$-space of the region occupied by any arbitrary number $n$ of atoms becomes more and more dispersed, whilst remaining of constant volume. As has been said, this 'spreading' has often been compared with the development of 'filaments' when a quantity of colour disperses in a stirred liquid, but this comparison can be deceptive since the latter process occurs in 'real' physical space.

The question arises how the volume $V_2$ can be brought into the picture as is required by equation (3.9). The answer put forward by those who advocate coarse graining is to use a 'smoothed out' density, $\bar{\rho}$, as described in §3.3.

It is much more physically realistic to accept that the Liouville theorem, in its preceding form, becomes less and less applicable as the 'filaments' in $\mu$- (or $\Gamma$-) space become finer and finer. This is because allowance must be made for the minute random disturbances which inevitably affect the system even though it is supposed to be 'isolated'. Roughly speaking, the randomising influences may be said to act on the filamentous structure in the phase space and to break it up. From a more formal viewpoint, the same variable influences imply the need for introducing into the Liouville equation some additional stochastic term such as was proposed by Bergman and Lebowitz. Thus the restriction (3.1) becomes progressively less and less applicable as the filaments become finer and finer. As a consequence there develops a genuinely homogeneous equilibrium distribution in phase space in place of the striated structure of the type shown in Figure 1(b). Equation (3.9) then follows quite naturally from the

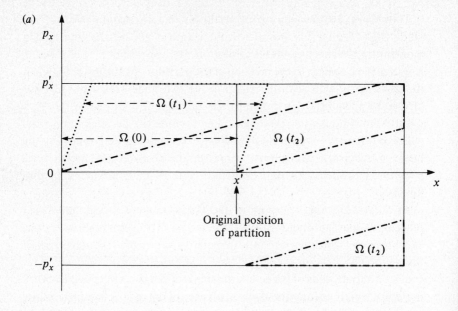

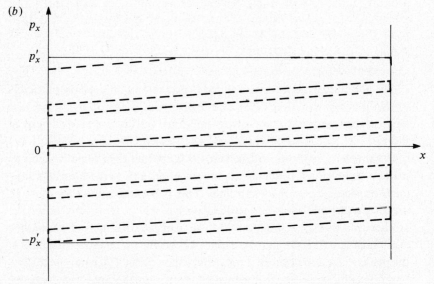

Figure 1. Expansion of a gas into a vacuum; two-dimensional phase space diagram with coordinates $x$ and $p_x$.

(a) $\Omega(0)$ refers to an arbitrary phase area when the gas was originally confined by the partition; $\Omega(t_1)$ and $\Omega(t_2)$ refer to the corresponding and equal phase areas at times $t_1$ and $t_2$ after removal of the partition.

(b) The area of $\Omega$ at a much later time $t_3$ showing the striated structure which continues permanently if Liouville's theorem remains strictly applicable.

statistical mechanics of equilibria by applying the corresponding theory to the change from the initial equilibrium state of volume $V_1$ to the final equilibrium state of volume $V_2$.

Thus, if coarse graining is deemed unacceptable, this example is sufficient to indicate that self-consistency in the theoretical structure of statistical mechanics requires the taking cognizance of externally induced perturbations.

Let us add what was already said at the end of §3.4. We do not regard the external influences as being the *cause* of the gas expanding – that expansion is due to the atoms being in motion. Similar considerations apply to other kinds of irreversible processes. But we do regard them as being the origin of the eventual establishment of an equilibrium condition to which the microcanonical or canonical distribution of probabilities can be properly applied.

### 3.6. Irreversible processes for which coarse graining can be dispensed with

As is well known, the thermodynamic treatment of the Second Law starts from an 'impossibility principle' which is very thoroughly supported by practical experience. Kelvin's form of the principle is as follows:

> It is impossible to take heat from a reservoir at a uniform temperature and convert it into mechanical work without leaving a change in the thermodynamic state of some other body.

It is, of course, the immense 'strength' of this principle which allows thermodynamics to proceed so easily to prove the existence of entropy as a function of state and also its increasing property. Statistical mechanics has no comparably 'strong' principle at its basis and indeed it denies the universality of the foregoing thermodynamic principle since it is shown to have a merely probabilistic character.

There are, of course, a truly vast variety of irreversible processes which fall within the scope of the thermodynamic Second Law – processes of mixing, friction, chemical reaction, radioactive decay and so on. Van Hove (1957) has expressed scepticism about whether statistical mechanics can ever be hoped to achieve the same degree of comprehensiveness. He remarked that no general set of equations is known which is capable of describing the non-equilibrium behaviour of many-particle systems 'and in view of the unlimited diversity of possible non-equilibrium situations, the existence of such a set of equations seems rather doubtful'. Van Hove's point has been greatly fortified by more recent studies (Prigogine, Berry and others) which have shown the prevalence of *instabilities of motion* even within the apparently deterministic situation of classical mechanics.

Thus it may well be the case that statistical mechanics has to tackle the 'problem of irreversibility' piecemeal, by using different theoretical methods as between one kind of situation and another.

Let us ask, then, whether there are any irreversible processes whose temporal development can be seen as a gradual unfolding of newly accessible 'states' whose probabilities are given throughout by the micro-canonical or canonical distributions and which can be accounted for by using only the fine grained entropy.

Such an approach seems out of the question in the foregoing example of free gas expansion, since as soon as a valve is opened between one vessel and another which is evacuated all sorts of eddies and surges develop. These are highly 'unknowable', especially if the vessels have an irregular shape.[29] Yet there may be other examples of irreversibility to which this notion of a gradual unfolding might be applied. A desirable condition would seem to be that the system in question remains homogeneous on the molecular scale, i.e. that spatial asymmetries and non-uniformities do not occur. Under such conditions it might be hoped that quite a small number of variables would be sufficient to specify the degree of change within the entire system at any moment. As we shall show below, a homogeneous chemical reaction in a gaseous or liquid medium is one such process.

### 3.7. The example of chemical reaction

There is, of course, an immense diversity of speeds among chemical reactions. Explosive reactions may complete themselves in a fraction of a second; other reactions such as the combination of hydrogen and oxygen at room temperature are so slow that no significant formation of water occurs even during geological periods of time. We shall here give attention to reactions, occurring within a homogeneous gaseous or liquid system, which are a good deal slower than the explosive ones.

There are three features of such reactions which make them promising candidates for what was proposed at the end of §3.6. These are:

(a) If the reagents are mixed together by rapid stirring, aided by the effect of molecular diffusion, the reaction can be regarded as actually starting, at a moment $t_0$, within a system which is already homogeneous.

(b) At times $t > t_0$ the chemist is able to make experimental measurements of the momentary composition of the system, together with its temperature and volume, which are very nearly as accurate as he would make at the eventual equilibrium state. Indeed, in the case of reactions which are catalysed, the reaction can be brought to a stop, at any chosen degree of conversion, by a sudden removal of the catalyst. It is considerations such as these which have convinced the chemist that a free energy, or an entropy, can be justifiably attributed to the reacting system at any stage of the reaction process.

(c) In the case of slow gaseous reactions there is good evidence that the canonical distribution is not significantly departed from, at any stage of reaction, as regards the translational, rotational and vibrational modes. This point has been established by Meixner, Prigogine and others, and it is particularly important in regard to the concept of temperature which otherwise would not be well defined. That temperature remains meaningful seems to apply even in the case of quite fast reactions, although perhaps not under such extreme conditions as occur in explosions or in the flame fronts of burning gases.

The assumption of the maintenance of the canonical distribution is implicit in the 'transition state theory' of chemical kinetics. Here it is supposed that reaction occurs via an 'activated complex', having an energy sufficient to pass over a potential energy barrier, and whose amount is so very slowly depleted by conversion into reaction product that its concentration remains essentially at equilibrium with the reagents. For example, in the case of a reaction $A + B \rightarrow C$, passing through the activated complex $X$, the scheme is taken to be

$$A + B \leftrightharpoons X \rightarrow C$$

where the curved arrows indicate processes taking place so fast, relative to the rate of conversion of $X$ into $C$, that $X$ remains substantially at equilibrium with $A$ and $B$. Similarly, the activated complex $X'$ which occurs in the reverse reaction is taken as being at equilibrium with $C$.

It may be added that it is precisely because the rates of reactions are determined by the height of an energy barrier, and by the presence of a minute fraction of molecules having energy sufficient to pass over the barrier, that the rates are singularly little affected by the entropy, or free energy, difference between reagents and products. Thus it can occur that very slow reactions may nevertheless be accompanied by a large increase of entropy, or by a large decrease of free energy.

Consider a reaction

$$A + B \rightleftharpoons C$$

and let $N_a$, $N_b$ and $N_c$ be the numbers of molecules of $A$, $B$ and $C$

respectively in the volume, $V$, of the container at any instant $t \geqslant t_0$. It is these numbers, together with the temperature and volume, which will be taken as determining the accessible quantum states at any instant of reaction. $N_a$, etc., are not coarse grained variables, of the type of $\bar{\rho}$ in equation (3.3), since they refer to the whole physical volume of the container and not to some 'star', of arbitrarily chosen size, in the $\Gamma$-space. They are, rather, thermo-dynamic variables of the same kind as temperature and volume.

Since the partition function is directly related to the Helmholtz free energy by equation (2.25),

$$\mathscr{A} = -kT \ln Q$$

it will be simplest to prove that the free energy is a decreasing function when the system is held at constant temperature and volume, rather than to prove the thermodynamically equivalent theorem that the entropy is an increasing function if the system were held at constant energy and volume. Again for simplicity we restrict ourselves to the case of a perfect gas mixture. The partition function then readily factorises as

$$Q = \frac{1}{N_a! \, N_b! \, N_c!} \cdot f_a^{N_a} f_b^{N_b} f_c^{N_c}$$

where $f_a$, etc., are the *molecular* partition functions and depend on molecular energy states (calculated relative to the same zero), rather than whole-system energy states such as we were concerned with in §2.4. This equation is proved in the standard texts, such as Fowler and Guggenheim (1939), and will here be taken for granted. It is based on the canonical distribution,[30] and, as has been said, this distribution will be assumed to hold during the reaction.

Using Stirling's approximation for the factorials, and noting the stoichiometric restrictions:

$$dN_c/dt = -dN_a/dt = -dN_b/dt$$

which apply to this reaction, we obtain from the preceding equations:

$$\frac{1}{kT} \, (\partial \mathscr{A}/\partial N_c)_{T,V} = \ln \left( \frac{N_c}{N_a N_b} \cdot \frac{f_a f_b}{f_c} \right)$$

Let $\bar{N}_a$, $\bar{N}_b$ and $\bar{N}_c$ be the number of molecules which satisfy the condition

$$\frac{\bar{N}_c}{\bar{N}_a \bar{N}_b} = \frac{f_c}{f_a f_b} \tag{3.10}$$

so that the previous equation can be re-written as

$$\frac{1}{kT} \, (\partial \mathscr{A}/\partial N_c)_{T,V} = \ln \left( \frac{N_c}{N_a N_b} \cdot \frac{\bar{N}_a \bar{N}_b}{\bar{N}_c} \right) \tag{3.11}$$

or as

$$\frac{1}{kT} (\partial \mathscr{A}/\partial t)_{T,V} = \ln \left( \frac{N_c}{N_a N_b} \cdot \frac{\bar{N}_a \bar{N}_b}{\bar{N}_c} \right) (\partial N_c/\partial t)_{T,V} \qquad (3.12)$$

Suppose that at $t_0$ the system contains $N_a^0$ molecules each of $A$ and $B$, together with $N_c^0$ molecules of $C$ for the sake of complete generality. Put $N_a^0 + N_c^0 = N^0$. Then from the stoichiometry of the reaction:

$$N_a = N_b = N^0 - N_c$$
$$\bar{N}_a = \bar{N}_b = N^0 - \bar{N}_c \qquad (3.13)$$

From (3.12) and (3.13) we obtain:

$$\frac{1}{kT} (\partial \mathscr{A}/\partial t)_{T,V} = \left[ \ln (N_c/\bar{N}_c) + 2 \ln \left( \frac{N^0 - \bar{N}_c}{N^0 - N_c} \right) \right] (\partial N_c/\partial t)_{T,V} \qquad (3.14)$$

Now, if $N_c$ is zero the reaction cannot occur in the direction $C \rightarrow A + B$. Also in the general case[31] the reaction will not be at equilibrium. Thus when $N_c$ is zero, or is sufficiently small, the reaction must be proceeding in the direction $A + B \rightarrow C$ and therefore $(\partial N_c/\partial t) > 0$. Similarly if $N_a$ and $N_b$ are zero, or are sufficiently small, the reaction must be proceeding in the direction $C \rightarrow A + B$, and therefore $(\partial N_c/\partial t) < 0$. This much can be said without taking any assumptions from the theory of chemical kinetics.

A state of equilibrium clearly requires $\partial N_c/\partial t = 0$, implying $\partial \mathscr{A}/\partial t = 0$. Suppose that during reaction $N_c$ has reached the value $\bar{N}_c$. Then from (3.13) and (3.11), $N_a = \bar{N}_a$, $N_b = \bar{N}_b$ and $\partial \mathscr{A}/\partial N_c = 0$, again implying $\partial \mathscr{A}/\partial t = 0$. Consistency with what was said in the previous paragraph is obtained if the turning point from $\partial N_c/\partial t > 0$ to $\partial N_c/\partial t < 0$ occurs when $N_c/\bar{N}_c$ changes from being $< 1$ to being $> 1$. This implies that the numbers $\bar{N}_a$, $\bar{N}_b$ and $\bar{N}_c$ are the numbers of molecules at equilibrium and (3.10) is the condition for equilibrium.

Now the logarithmic terms in (3.14) are both negative when $N_c/\bar{N}_c < 1$, and, as has just been said, $\partial N_c/\partial t$ is then positive. Therefore $\partial \mathscr{A}/\partial t < 0$. Similarly, the logarithmic terms are both positive when $N_c/\bar{N}_c > 1$, and simultaneously $\partial N_c/\partial t$ is negative. Thus $\partial \mathscr{A}/\partial t$ is again negative. In short, $(\partial \mathscr{A}/\partial t)_{T,V} < 0$ whether reaction is proceeding in the one direction or the reverse, and the Helmholtz free energy has a minimum value at equilibrium, a result already familiar from thermodynamics.

In this example of chemical reaction in a perfect gas mixture it has thus been possible to argue, without recourse to coarse graining, that the Helmholtz free energy diminishes continuously throughout the approach to equilibrium at constant temperature and volume. This is a much stronger conclusion than could be drawn from the $H$-theorem, although the latter, of course, is much more general.

Our objective above has been a modest one; it has been simply to show, by means of a counter-example, that coarse graining, with its possibly subjective aspect as a 'loss of information', is by no means *essential* (as van Kampen, 1962, and others, have maintained) to the treatment of irreversibility.

Although there are many situations where coarse graining may offer the best approach which is presently available, it is surely to be regarded as providing nothing more than a *model* of the real process – a model which may well be inadequate. Thus any subjectivity which is implied by this particular model need not necessarily be 'read into' the actual process. As is well known, methods for avoiding coarse graining which are much more elaborate and comprehensive than we have adopted above are in process of development by Prigogine (1980) and others.

### 3.8. Review

Our first three chapters have led to the conclusion that the charges laid against entropy, as summarised in §1.1, are by no means justified. Let us ask why so many scientists from Maxwell onward have held a contrary opinion and have believed entropy to be a subjective concept.

Some of the reasons have been put forward already. Maxwell's notion that entropy is related to disorder, and if so is mind-dependent, was discussed in §1.1 and §2.5. Another factor was brought up in §1.3. This was the tendency to overlook the epistemological difference between the real *properties* of substances, on the one hand, and the *values* attributed to the properties on the other. Although the latter depend on the state of knowledge, this does not impugn the objectivity of the properties. As for Gibbs' Paradox, which Maxwell saw as showing the subjectivity of entropy, this will be taken up in Chapter 4. It will be shown that the Paradox can be resolved without invoking human powers of knowing.

An interesting point was made by Hobson (1971). Entropy, he suggested, was regarded as being subjective because it is not a *mechanical* quantity and because there was an established view in science to the effect that only mechanical quantities are, in some sense, real. Since entropy is a *dispositional* property of matter in mass, and cannot be dealt with in terms of the motions of single atoms, it seemed to those who held the traditional mechanical concepts of science that entropy must be unreal.

Popper (1982), we think, comes even closer to the heart of the matter. He draws attention to the fact that scientists, throughout most of the history of science, have cherished a belief in the complete determinism of natural phenomena and have therefore denied that there is any objective chance or randomness. This belief in determinism, he says, is incompatible with objective physical probabilities.[32] It followed, on this basis, that any apparent randomness in natural phenomena can be nothing more than a manifestation of human ignorance, and does not inhere in the phenomena themselves. As an example of this outlook he instances Einstein's deeply held conviction that any probabilistic theory in physics cannot be regarded as fully objective and complete. Popper goes on to deal with the implications of a belief in determinism for the statistical interpretation of thermodynamics: it meant that the probabilities involved in this interpretation were necessarily *subjective probabilities* which reflected incomplete knowledge. Therefore the entropy which emerged from the statistical mechanical theory was also regarded as subjective.[33]

Of course, certain attitudes have changed. In particular, the difficulties encountered by hidden variable theories in quantum mechanics have made the belief in a basic determinism seem very insecure.

However, a further influence became manifest in the '60s from a quite different direction, unrelated to the question of determinism. This was the development of information theory and its application to a wide variety of statistical problems. Although its *formalism* is undoubtedly more comprehensive than is that of statistical mechanics, its *interpretative language* has been restricted by the assumption, in Jaynes and others, that the probabilities it refers to are always subjective probabilities. As a consequence those who came to regard statistical mechanics as being a special case of information theory introduced terms such as 'ignorance' and 'surprisal' which implied a very significant epistemological standpoint. No doubt such terms were appropriate to many of the applications of information theory, but they seem entirely alien to the discussion of physical phenomena. Accordingly we shall show in Chapter 5 that statistical mechanics remains a self-sufficient discipline and does not require the kind of interpretative language which has been imported into it by some of the information theorists.

In the meantime, three points are worth re-emphasising: (1) The originally defined thermodynamic entropy was seen in Chapter 1 to be fully objective, and the statistical mechanical treatment has been shown in Chapters 2 and 3 to do nothing to undermine that conclusion apart from the controversial matter of coarse graining; (2) As was argued at some length in §1.2, incompleteness of knowledge is not a sufficient reason for the imputation of subjectivity in what that knowledge refers to; (3) Although it is undoubtedly a fact that 'we cannot know' the exact coordinates and momenta of molecules, this is nevertheless *an entirely adventitious fact*; as was said in §2.5, it has no bearing on the fully objective character of irreversible phenomena such as the escape of a gas from an unstoppered bottle, or the equalisation of temperatures between two bodies. Although it may be conceded that entropy can be regarded, if one so wishes it, as a measure of 'missing information', this is irrelevant to the physics of the situation. No amount of information about the coordinates and velocities of molecules would prevent an irreversible process from occurring – so long, at least, as the recipient of that information is entirely passive, or is unable to use it to act on the molecules.

# 4

## *Identity and indistinguishability*

### 4.1. Introduction

Two developments which gave rise to great controversy were the discovery of the entropy of mixing paradox (Gibbs, 1876) and the realisation that statistical mechanics, as it was in the early 20th century, resulted in a non-additive entropy. As a consequence, scientists were obliged, perhaps for the first time, to give serious thought to the philosophical concepts of identity and indistinguishability. It had to be asked, for example, why a properly additive entropy[34] seemed to require that atomic particles of the same kind must be regarded as indistinguishable whereas macroscopic bodies, such as billiard balls, may be readily individuated however alike they may be. Questions had to be asked about the criteria of identity and indistinguishability, and about whether or not there is a continuous passage from indistinguishability to distinguishability in the transition from the microscopic to the macroscopic domains.

A possible answer to these questions is that it is a matter of human powers of discrimination. We can locate billiard balls and follow their individual motions, but we cannot do the same with electrons. If this were the correct answer it would imply that distinguishability is a subjective matter, but this we shall seek to refute.

An alternative answer which has sometimes been put forward is that Heisenberg's Principle is at the root of the apparent difference between atomic indistinguishability, on the one hand, and the individuation of macroscopic entities, on the other. Now it is true, of course, that when two 'identical' particles collide it is not possible to say whether the cloud chamber track emerging to the left corresponds to 'the same' particle as that which gave rise to the left-hand track before collision. But it is also true that the notion of indistinguishability had been accepted within classical statistical mechanics, in order to obtain an additive entropy, long before Heisenberg's Principle was dreamed of. In short, there was an *ad hoc* justification for it in advance of quantum theory.

It will be useful, therefore, briefly to trace the causes of a great reversal of opinion – from the 19th century view that it is natural to regard atomic particles, even those which are of the same kind, as being capable of being 'individuated', to the modern view that such particles cannot be individuated. (To individuate: to distinguish from others of the same species. (Oxford Dictionary.)) This change of attitude is not sufficient, however, to resolve Gibbs' Paradox. The latter, to which we turn in §4.3, is a macroscopic phenomenon and is not concerned with particles but with macroscopic quantities of material. Our treatment of the Paradox will be followed in §4.6 with a somewhat philosophical discussion on identity and indistinguishability. After that we give a brief account of the symmetry rules and then turn finally to the question, as raised above, about whether there is continuity between the micro- and macro-levels in regard to the matter at issue.

## 4.2. The additivity of the entropy of gases

Although in Chapter 2 we outlined the route into statistical mechanics via the partition function, an earlier route, as used by Boltzmann, depended on combinatorial methods. Let us, therefore, first say a few words about 'Boltzmann counting' before coming back to the use of the partition function.

In his simplest example of the combinatorial method, Boltzmann (1877) considered a gas of $N$ 'like' particles and he supposed that the energies of the particles could be regarded as lying within discrete ranges such that $n_0$ of them momentarily have zero energy, $n_1$ have energy $\varepsilon$, $n_2$ have energy $2\varepsilon$, and so on. (This was not an anticipation of quantum theory and was adopted only for simplicity.) He then defined what he called the 'permutability' $P$:

$$P \equiv N!/n_0!\, n_1!\ldots \tag{4.1}$$

Now this expression will be recognised as being the number of ways in which $N$ *distinguishable* entities can be arranged in piles or boxes: $n_0$ in the first pile, $n_1$ in the second pile, and so on. Although Boltzmann was dealing with 'like' particles he thought of them as being individually identifiable. This was a natural enough assumption at his time since classical mechanics certainly regarded macroscopic bodies as being individuals and there seemed no good reason, within the scope of the classical theory, why individuality should go over into non-individuality as particles became smaller and approached the atomic in size. Boltzmann spoke more specifically about gas molecules as being traceable by their initial conditions and by the continuity of their motions. This condition, he said, 'gives us the sole possibility of recognising the same material point at different times'. (Quoted by Pais, 1979.)

Boltzmann proceeded to find the distribution over the energy states which maximised the logarithm of $P$, subject to the sum of the $n_i$ being equal to $N$ and subject also to the sum of the terms $n_i \varepsilon_i$ being equal to the total energy of the system. All individual 'complexions' having the given total energy were assumed to be equally probable and thus the maximisation of $\ln P$ yielded 'the most probable distribution', i.e. the one comprising the largest number of complexions or microstates. Using Stirling's approximation and the method of undetermined multipliers, he was thus able to show that the combinatorial method yielded the same Maxwell–Boltzmann distribution, $n_i \propto \exp - \beta \varepsilon_i$, as had been obtained previously by other methods of derivation. However, the same combinatorial method *did not provide* an additive expression for the entropy. To achieve that would have required making an *ad hoc* division of $P$ by $N!$. The effect of doing this, making $P \leqslant 1$, would clearly have made nonsense of the meaning of $P$ as a 'number of ways'.

The history of the combinatorial method became even more involved during the early quantum period which was marked by intense controversy between Planck and Ehrenfest. One of the issues was whether $S_{BP}$ should be written as $S = k \ln W$ or as $S = k \ln W + A$ where $A$ might be a function of $N$. Planck's inclination was to eliminate $A$ and thereby to adopt an 'absolute' entropy, as was suggested to him by Nernst's Heat Theorem which led on to the Third Law. It was also appreciated that Stirling's approximation was not applicable unless the $n_i$ were large enough numbers. This was one of the reasons for the adopting of 'cells' in the $\mu$-space such that many quantised energy levels were comprised within each energy cell. In place of $P$ one wrote:

$$W = \frac{N!}{n_1! \, n_2! \ldots} \, \omega_1^{n_1} \omega_2^{n_2} \ldots \tag{4.2}$$

where the term $\omega_i^{n_i}$ is the number of ways in which $n_i$ independent and distinguishable molecules can be put into the $i$th cell which comprises $\omega_i$ energy levels. Maximisation of $W$ gave 'the most probable distribution' within the coarse grained system of cells and this maximum value of $W$ was used in conjunction with $S_{BP} = k \ln W$ to calculate the entropy. What is very relevant to the present discussion is that this procedure again failed to produce an additive entropy except when $N!$ was removed from the numerator of (4.2). The result of dividing (4.1) or (4.2) by $N!$ constitutes what is called 'corrected Boltzmann counting'. Yet this was widely regarded as being not at all well justified until wave mechanics brought a new approach to the problem in 1924. The history of these difficulties is well told by Klein (1958), Hanle (1977) and Kastler (1983).

Let us turn to Gibbs and to the method of approach based on the partition function which, in our view, brings out the essential point much more clearly than can be achieved by the combinatorial method of Boltzmann.

Gibbs was the true originator of the Indistinguishability Postulate for in his 1902 book (Dover reprint 1960, p. 187) he remarked that 'it seems in accordance with the spirit of the statistical method' to regard the interchange of 'entirely similar particles' as not giving rise to a changed state of the overall system. He went on to say that this proposal was in no way contrary to the idea that particular particles can be identified even when they are of the same kind. In the light of both of these remarks, it seems that Gibbs drew a distinction between the indistinguishability of like particles when the matter is considered in a statistical context, on the one hand, and the supposed distinguishability of the same particles in the context of exact mechanics, on the other. His final paragraph in the chapter, when taken in conjunction with his equation 546, makes it clear that the correct entropy, in his view, requires the division of the classical partition function (as it was later called) by $N!$. He may have reasoned that since the entropy, like the Hamiltonian itself, is a property of *the whole* gas, the interchange of like particles should have no statistical consequence.

For the purpose of investigating the matter quantitatively let us consider a dilute monatomic gas consisting of $N$ particles enclosed in a volume $V$ at a temperature $T$. The classical analogue of the partition function (2.17) is obtained by replacing the summation in that equation by an integral over the coordinates and momenta. Thus

$$Q_{class.} = \int \ldots \int \exp(-H/kT) \, dq_1 \ldots dq_{3N} \, dp_1 \ldots dp_{3N}$$

where $H$ is the classical Hamiltonian for free mass points of mass $m$:

$$H = \sum_{i=1}^{3N} p_i^2/2m$$

The classical partition function is thus readily evaluated as

$$Q_{class.} = V^N (CmT)^{3N/2}$$

where $C$ is a constant $(C = 2\pi k/h^2)$ which is independent of the nature of the atoms.[35] If we now apply equation (2.24), the entropy of the gas is obtained as

$$S' = kN\{\ln V + \tfrac{3}{2}\ln(CmT) + \tfrac{3}{2}\} \qquad (4.3)$$

Now this expression is unsatisfactory since if we consider a gas twice as large, with $N$ and $V$ both doubled, $S'$ is not doubled. As with Boltzmann's combinatorial method, the 'entropy' which is obtained is not additive. Indeed, suppose we have two samples of the same gas, each containing $N$ atoms in a volume $V$, and remove a partition between them at constant temperature. The formula gives

$$\Delta S' = 2kN \ln 2V - 2kN \ln V = 2kN \ln 2$$

However, this should surely be zero since, as Schrödinger put it, no real event occurs when a gas diffuses into itself.

If, in accordance with Gibbs' proposal, we divide $Q$ by $N!$ we obtain for the 'corrected' partition function:

$$Q_{\text{corr.}} = V^N (CmT)^{3N/2}/N! \tag{4.4}$$

Using Stirling's approximation,[36] $\ln N! \approx N \ln N - N$, the entropy becomes:

$$S = kN\{\ln (V/N) + \tfrac{3}{2} \ln (CmT) + \tfrac{5}{2}\} \tag{4.5}$$

This is the Sackur–Tetrode equation, first obtained in 1911–13, and it is seen to be entirely satisfactory in regard to the additivity of entropy. For example, for the foregoing process of two equal portions of the same gas 'mixing' with themselves:

$$\Delta S = 2kN \ln (2V/2N) - 2kN \ln (V/N) = 0 \tag{4.6}$$

The division by $N!$ has thus had the effect of providing an entropy for monatomic gases which is consistent with thermodynamics. Yet the division did violence to the classical view that individual atoms are capable, at least in principle, of being distinguished from each other by their trajectories. As Sterne (1949) put it: 'The conception of atoms as particles losing their identity cannot be introduced into the classical theory without contradiction.'

With the coming of modern quantum mechanics, from 1925 onwards, new ideas were brought to bear on the problem. Dirac advanced the postulate that no *observable* change occurs in a system consisting of particles of the same kind when any two of them are interchanged. This Indistinguishability Postulate, as it is called, implies that even when the particles are mechanically independent (e.g. in a dilute gas) they are not statistically independent.

Many textbooks present the quantum derivation of equation (4.5) as if the Postulate immediately justifies the former *ad hoc* division by $N!$. In our view a much more satisfactory derivation was that of Schrödinger (1948, p. 56), who obtained the equation without making that division and without adopting the Stirling approximation. He used the Darwin and Fowler method for evaluating the partition function of an assembly of bosons or fermions and showed that, under the conditions of a dilute monatomic gas at not too low a temperature, the entropy is exactly as is given by (4.5).

Further support for this equation is obtained by comparing its numerical evaluation with calorimetric entropies based on the Third Law. For example, the monatomic vapour of mercury, at 343.9 K and at its vapour pressure, has an entropy value of 257.3 J K$^{-1}$ mol$^{-1}$ by equation (4.5). The calorimetric value is 256.1 as obtained by integrating d$q/T$ from close to the absolute zero using measured specific heats and latent heats of phase change (Denbigh, 1981 *a*, p. 427 and p. 486). Rowlinson (1963, p. 31) reports similarly close agreement in the cases of argon and thallium vapour which are also monatomic. As was remarked in Chapter 1, this degree of agreement between the results of two independent theories also gives strong support to the view that entropy is fully objective.

### 4.3. Gibbs' Paradox

This Paradox is presented in some of the textbooks in a rather confusing way, as if it has two different forms. It is important to realise that the matter of obtaining an additive entropy within statistical mechanics, as dealt with above, *does not solve* the Paradox in the form in which it was originally put forward by Gibbs (1876). He obtained it on the basis of a purely thermodynamic – and therefore macroscopic – process of reasoning.

Nevertheless it will be useful, before turning to the thermodynamics, to show how the Paradox also arises within the statistical mechanical context of the previous section. It is a matter of now attending to *mixtures*. For simplicity consider a mixture of two monatomic gases, $A$ and $B$, containing $N_a$ atoms of the one component and $N_b$ atoms of the other. The Sackur–Tetrode equation for this mixture, analogous to (4.5) for each gas singly, can be derived as:

$$S_{ab} = kN\{\ln V + \tfrac{3}{2} \ln (CmT) + \tfrac{5}{2}\} - k(N_a \ln N_a + N_b \ln N_b) \qquad (4.7)$$

where $N = N_a + N_b$ and $m^N = m_a^{N_a} m_b^{N_b}$. The entropy increase on the mixing of the pure gases $A$ and $B$, having volumes $V_a$ and $V_b$ respectively such that $V_a + V_b = V$ (i.e. mixing at constant total volume), is readily obtained from (4.5) and (4.7) and is

$$\Delta S_{\text{mix.}} = S_{ab} - S_a - S_b = -kN_a \ln (V_a/V) - kN_b \ln (V_b/V)$$
$$= -kN_a \ln x_a - kN_b \ln x_b \qquad (4.8)$$

The last line follows from the gas laws $pV_a = n_a RT$ and $pV_b = n_b RT$. $x_a$ and $x_b$ are the mole fractions $N_a/N$ and $N_b/N$ respectively. By applying the Avogadro number, $L$, (4.8) can be transformed to the alternative form:

$$\Delta S_{mix.} = -R(n_a \ln x_a + n_b \ln x_b) \qquad (4.9)$$

where $n_a$ and $n_b$ are the numbers of moles of $A$ and $B$ respectively.

Now this expression contains no variable which allows for the degree of difference between $A$ and $B$. Gibbs, who had obtained the same equation from thermodynamics, put the point as follows: the entropy of mixing of different gases (at constant temperature and volume) is 'independent of the degree of similarity or dissimilarity between them', but is *zero* when two samples of *the same* gas are mixed. It was this strange discontinuity, i.e. that the entropy of mixing *remains constant* however alike the gases are, and then suddenly collapses to zero when the gases become 'the same', which constitutes Gibbs' Paradox.

It will be clear that the quantal treatment, via the Indistinguishability Postulate, of the additivity of entropy in no way resolves this Paradox. The indistinguishability there referred to is about the *particles* of some particular kind, but the Paradox is about matter in bulk in the form of mixtures. Unless it can be resolved it would appear that the 'identity' of bulk matter is not the limit of its vanishing indistinguishability.

In his thermodynamic derivation of (4.9) Gibbs used his newly developed method of chemical potentials. Some twenty years later Planck obtained the same equation by a different thermodynamic argument involving the separation of mixed gases through semi-permeable membranes. Guggenheim[37] remarked that this had some advantage over the Gibbsian method. On the other hand, Planck recognised that semi-permeable membranes can never be perfect, as a matter of principle, and this brings into question whether his own method of derivation was fully self-consistent.

Let us examine this point, since it has a bearing on our later resolution of the Paradox. What is meant by a semi-permeable membrane is a membrane which permits the free passage of *only one* component of a mixture – to all other components it is taken as being entirely impermeable. It is supposed that there is equality of the chemical potentials across the membrane for the one component but not for the rest, and this is the essential origin, in the Planck derivation, of the discontinuity in the entropy of mixing. Yet Planck (1927, p. 210 and p. 238) admitted that when the components are sufficiently similar this condition can no longer hold since their diffusion rates through the membrane then become comparable in magnitude. Even so, he did not regard this as invalidating his derivation of

(4.9), since he took this equation as representing an idealised limiting case which was justified by the fact that there exist certain membranes which approximate very closely to the foregoing criterion of semi-permeability relative to certain substances. He quotes (1927, p. 218) the example of the separation of hydrogen from air by use of platinum foil at a white heat; the foil is readily permeable to hydrogen but not measurably so to air. Palladium is now known to be even more effective as a means of preparing very pure hydrogen.

Let us leave this aside for the moment and return to the Paradox. For a total of $\mathcal{N}$ components the general expression for the entropy of mixing at constant volume and temperature is

$$\Delta S_{\text{mix.}} = -R \sum_{i=1}^{\mathcal{N}} n_i \ln x_i \qquad (4.10)$$

and the thermodynamic argument shows this to be applicable to liquid and solid solutions, as well as to gaseous mixtures, so long as the solutions are 'ideal' over the whole of their composition range. (See Appendix 4.1.)

Now (4.10) is clearly a step function in regard to the number of components. Any reduction of $\mathcal{N}$ gives rise to a discontinuous reduction of $\Delta S_{\text{mix.}}$ and in particular when $\mathcal{N}$ goes to unity $\Delta S_{\text{mix.}}$ falls to zero. However, this point does not bring out what is most peculiar about the equation. For simplicity, consider the mixing of one mole each of only two components. Then we obtain the simple numerical value

$$\Delta S_{\text{mix.}} = 2R \ln 2 \qquad (4.11)$$

Thus what is so striking is that, as the two components are made more and more similar, the entropy of mixing *remains constant* at the value $2R \ln 2$. For instance, the entropy of mixing of ortho- and para-hydrogen, which differ from each other only in regard to their nuclear spins, is apparently no smaller than is the entropy of mixing of, say, helium and carbon dioxide. It is the absence of any 'warning' that the entropy change will suddenly collapse to zero, when the substances are 'the same', which is the truly paradoxical aspect of the matter.

Since the foregoing equations are obtainable from thermodynamics, as well as from the quantal statistical mechanics, it would appear, as mentioned already, that the Paradox is not directly related to the indistinguishability of atoms or molecules of the same kind. Thermodynamics is a science which makes no reference whatsoever to the existence of such entities! It deals entirely with macroscopic concepts, such as heat and mechanical work, energy, entropy and temperature, and it would seem therefore that the 'indistinguishability' and the 'identity' we are here concerned with is that of macroscopic quantities of material. Also, since mixing is a macroscopic process, it might be expected that a resolution could be achieved by using the resources of thermodynamics alone.

Before attempting this in the following section let us give a brief summary of the extensive literature.[38] But first it should be said that several authors, writing from the viewpoint of statistical mechanics rather than from that of thermodynamics, seem not to take account of the fact that the Paradox applies as strongly to ideal liquid and solid solutions as it does to ideal gaseous mixtures.

A number of scientists (e.g. Fast, 1970; Mandl, 1974) have maintained that the identity of a substance with itself is not something which can be approached gradually; either self-identity holds or it does not, and this (they say) is the source of the discontinuity in the entropy. This answer is unsatisfying, since it does not explain why there is not at least a *diminution* in the entropy change, as there is of other properties of mixtures, as the components are made more and more similar. Indeed, we normally expect that an infinitesimal change in one thing will not give rise to a discontinuous and finite change in something else.

A subjective view, favoured by Grad (1961) and by some of the information theorists (e.g. Hobson, 1971), is that the Paradox is a matter of a discontinuous change in the state of human knowledge regarding the number of substances which are mixed. (In information theory entropy is claimed to be an inverse measure of 'knowledge'.) Against this position we shall point out shortly that the entropy of mixing can be made physically manifest as a measurable quantity of heat, or of mechanical work, and, as such, is fully objective.

Other scientists (e.g. Kubo, 1965; Reif, 1965) have relied on the atomic structure of matter and have remarked that nature does not provide the means by which two substances can be made only infinitesimally different, due to their constituent particles differing by finite amounts of rest-mass, charge or spin. This answer, we think, evades the issue, since again, as above, it does not explain why two such very nearly similar substances as ortho- and para-hydrogen, to take a previous example, have precisely the same entropy of mixing as substances which differ greatly.

In our opinion, Gibbs himself (1876) offered the best clue to the resolution of his Paradox by anticipating the operational outlook in the philosophy of science. From the thermodynamic standpoint, he said, the issue concerns *separability*: when two samples of a gas are 'of the same kind' we are dealing with a situation where 'the separation of the gases is entirely impossible'. This is the idea we shall develop below. A mixture can, in fact, be shown to be a mixture only by separating its components, or by separating their manifestations. Similarly a substance can be said to be *pure* only if it is found to be non-separable. And of course these experimental procedures are fully objective.

### 4.4. A resolution of the Paradox

A mixing discontinuity, such as has been discussed, does not occur with properties such as volume, density, refractive index, etc. Why does the 'catastrophe' only occur in the case of entropy?

The explanation provides a further clue. It is due to (a) the dispositional character of entropy (see §2.6), and (b) the non-conservation of entropy except in idealised reversible processes. Any entropy change $\Delta S$ ($= S_2 - S_1$) between assigned states 1 and 2 of a system is not directly observable in itself but is a measure of the *disposition* of the system to give rise to a *maximal* heat effect in an external heat reservoir. But this upper bound is actually attained only when the process takes place along a path which is only infinitesimally displaced from equilibrium. Otherwise the change $\Delta S$, although remaining the same for the system in question (since entropy is a function of state), will manifest itself in an external heat effect which is *smaller* in magnitude. Indeed, if the process were to be carried out in an adiabatically isolated system the external effect would be zero. Thus, whereas in a reversible process the entropy change of the environment is equal in magnitude and opposite in sign to that of the system, in an irreversible process this equality no longer holds; there is a net entropy creation which yet may be unobservable externally.

Consider the spontaneous mixing of one mole each of two substances by the removal of a partition. We shall not question the correctness of equation (4.11), i.e. that the entropy creation is $2R \ln 2$, but would be zero if the substances were the same. Yet these calculated quantities are, in a sense, mathematical artefacts under circumstances where they do not manifest themselves as anything outside the system. In order to show them as physically real, the same mixing must be carried out reversibly so that the observable heat effect becomes maximal and definite in amount. Or, alternatively, one could look for a reversible separation since $\Delta S_{\text{mix.}} = -\Delta S_{\text{sep.}}$.

For these purposes we shall put forward a very reasonable postulate – namely that it is impossible to achieve an entirely complete separation of a homogeneous mixture in a single stage. A finite number of stages may be sufficient to attain an *acceptable* degree of separation – but this number will rise indefinitely if the further condition is imposed that the separation shall be reversible. Similarly, it is impossible to mix pure substances in a reversible manner and obtain a homogeneous mixture in a single stage. (From the thermodynamic viewpoint this follows from the fact that the chemical potentials of substances $B$, $C$, etc., in a pure substance $A$ are all minus infinity as will be seen from equations in Appendix 4.1.)

Much evidence is available in support of this postulate. Consider separation first. This requires a device, which may be called a *filter*, capable of discriminating between the substances present in the mixture. Typical devices are those which are used in fractional distillation (where the 'filter' is the liquid–vapour interface) and in selective diffusion through barriers. None of them are capable of achieving *complete* separation in a single 'stage' – the term we use for the period of time during which the separated portions of material remain in physical contact with each other, thus allowing transfer to occur.

For example, it will not be the case in a distillation process that all but one of the components of a liquid mixture are absolutely involatile. However small is the volatility of some of the components it will still be sufficient to contaminate the distillate with a few molecules of those components. Further purification can be achieved by resorting to a second, a third, and further stages of distillation, each of them a separate operation.

To be sure, what may appear as being a counter-example to the foregoing postulate is the almost perfect separation of hydrogen from other gases by passage through hot platinum or palladium foil. Yet this kind of process does not satisfy our second condition, namely reversibility. As has been said, 'semi-permeability' supposes equality of the chemical potentials across the membrane for *one* of the components; but not for the rest, contrary to the requirements for reversibility. No doubt such a supposition is a valid trick if the diffusion rates differ immensely. For one can then adopt the idealisation about some states being entirely 'accessible' whilst others are entirely 'inaccessible', as discussed towards the end of §2.3. Yet this becomes more and more unreal as the components become more and more similar – and this is the situation we are concerned with in the Gibbs' Paradox. One has only to think of the vast number of successive enrichment stages which are needed to obtain any appreciable separation of $U^{238}F_6$ and $U^{235}F_6$.

In short, separation processes are intrinsically 'statistical' in their action – the passing or the non-passing of individual molecules of types $A, B, C$, etc., through an interface is a probabilistic matter at any instant. As a consequence it will always occur that some undesired molecules will pass through and this becomes a very significant effect when the molecules are such that they differ only very slightly in their properties. The requirement of reversibility further reinforces what has been said since it necessitates that the material on the two sides of the interface must depart only infinitesimally from the state in which the chemical potential of each component is equal on each side. This is not compatible with there being *zero* concentrations of some components on the one side of the interface if there are finite concentrations on the other side.

That this is so follows from the fact that as the mole fraction of a component approaches zero its chemical potential approaches minus infinity. (See equations (1) and (3) of Appendix 4.1.) Thus a fluid in which the mole fraction of component $A$, say, approaches zero can only be in equilibrium, across an interface, with another fluid in which the mole fraction of that substance also approaches zero. Reversible separations to give *entirely pure* products thus require an infinite number of stages.

As regards mixing, there is this important difference – that the mixing of fully miscible substances can occur spontaneously and irreversibly. But as soon as *reversible* mixing is stipulated the same considerations apply as for a separation. The process must then be controlled in such a way that the portions of material which are being mixed in any stage never depart more than infinitesimally from equilibrium with each other. Thus here again an infinite number of stages are involved, in the reverse temporal sequence to the case of the corresponding separation.

Our proposal for the understanding of the Paradox will now be clear. The entropy of equation (4.10) as applied to a spontaneous and irreversible mixing is not physically real in the sense that such a process, *qua* being irreversible, does not allow its entropy change to become physically manifest as an observable and maximal amount of heat or of mechanical work. Only reversible changes give rise to such a manifestation and yet only *partial* mixings or separations can actually occur under reversible conditions. Thus equation (4.10) is essentially *the sum* of the externally observable entropy changes occurring in an infinite sequence of reversible stages, and *in any one* of them the entropy change varies continuously with the distinguishability of the components in that stage. In short, there is *complete continuity* in regard to what is observable. The apparent discontinuity of the Paradox could only be made manifest at the unattainable limit of an infinite number of reversible stages.

Let us substantiate this idea with an actual calculation. For purposes of simplicity this will refer to *the first* stage of a separation – or equivalently to *the final* stage of a mixing. Since the equations to be obtained below relate to a given physico–chemical system changing between defined initial and final states, it is not of immediate consequence whether the separation or mixing stage is regarded as occurring reversibly or not. This issue will be come to later.

Again for simplicity let the stage in question be the separation of an ideal mixture of only two substances, $A$ and $B$, one mole of each. The separation is necessarily incomplete and it yields two fractions, $\alpha$ and $\beta$, such that $\alpha$ is an $A$-rich fraction and $\beta$ is a $B$-rich fraction. Thus:

Mixture of one mole of $A$ and one mole of $B$

$$\rightarrow \begin{cases} \text{Fraction } \alpha: \ n_a^\alpha \text{ moles of } A \text{ and } n_b^\alpha \text{ moles of } B \\ \text{Fraction } \beta: \ n_a^\beta \text{ moles of } A \text{ and } n_b^\beta \text{ moles of } B \end{cases}$$

where the superscripts denote the fractions. From mass balance

$$n_a^\alpha + n_a^\beta = 1; \quad n_b^\alpha + n_b^\beta = 1 \tag{4.12}$$

Let $\delta S_{\text{sep.}1}$ be the entropy of this partial separation in stage 1. Let $\sum_{i=2}^{\infty} \delta S_{\text{sep.}i}$ be the sum of the entropy changes in all further partial separation stages required to complete the obtaining of pure $A$ and pure $B$. Having obtained these pure substances, let them be allowed to mix spontaneously and irreversibly. This results in the re-creation of the original mixture with entropy increase $\Delta S_{\text{mix.}}^{\text{total}}$, as given by equation (4.11). Since entropy is a function of state, it follows that

$$\delta S_{\text{sep.}1} + \sum_{i=2}^{\infty} \delta S_{\text{sep.}i} = -\Delta S_{\text{mix.}}^{\text{total}} \tag{4.13}$$

Now $\sum_{i=2}^{\infty} \delta S_{\text{sep.}i}$ is equal, again because entropy is a function of state, to the negative of the sum of $\Delta S_{\text{mix.}}^\alpha$ and $\Delta S_{\text{mix.}}^\beta$ where these two terms refer respectively to the entropies of the spontaneous remixing of pure $A$ and pure $B$ to form the separate fractions $\alpha$ and $\beta$. It follows that (4.13) can be replaced by:

$$\delta S_{\text{sep.}1} = -\Delta S_{\text{mix.}}^{\text{total}} + \Delta S_{\text{mix.}}^\alpha + \Delta S_{\text{mix.}}^\beta$$

The three terms on the right-hand side are readily calculable from (4.10) and are:

$$\Delta S_{\text{mix.}}^{\text{total}} = 2R \ln 2$$
$$\Delta S_{\text{mix.}}^\alpha = -R\{n_a^\alpha \ln x_a^\alpha + n_b^\alpha \ln x_b^\alpha\}$$
$$\Delta S_{\text{mix.}}^\beta = -R\{n_a^\beta \ln x_a^\beta + n_b^\beta \ln x_b^\beta\}$$

Thus we finally obtain for the first-stage entropy change:

$$\delta S_{\text{sep.}1} = -2R \ln 2 - R\{n_a^\alpha \ln x_a^\alpha + n_b^\alpha \ln x_b^\alpha + n_a^\beta \ln x_a^\beta + n_b^\beta \ln x_b^\beta\}$$
$$= -2R \ln 2 - R\{n_a^\alpha \ln x_a^\alpha + (1-n_a^\alpha) \ln (1-x_a^\alpha)$$
$$+ n_a^\beta \ln x_a^\beta + (1-n_a^\beta) \ln (1-x_a^\beta)\} \tag{4.14}$$

as follows from (4.12).

Consider two extreme cases:

(a) No separation is achieved in the first stage so that the fractions $\alpha$ and $\beta$ have the same composition and thus $x_a^\alpha = x_a^\beta = \frac{1}{2}$. Equation (4.14) then gives $\delta S_{\text{sep.}1} = 0$, as was to be expected.

(b) Almost complete separation is approached in the first stage so that $x_a^\alpha \rightarrow 1$ and $x_a^\beta \rightarrow 0$. The equation then gives $\delta S_{\text{sep.}1} \rightarrow -2R \ln 2$, as was also to be expected.

Between these upper and lower bounds the equation shows that $\delta S_{\text{sep.}1}$ *varies continuously* according to the extent to which $x_a^\alpha$ and $x_a^\beta$ differ – and this must clearly be determined by the degree of dissimilarity of the substances $A$ and $B$ in regard to that property of theirs on which the particular separation device operates. Furthermore, if the separation into the fractions $\alpha$ and $\beta$ can be carried out reversibly (and this may be a big 'if'!), $\delta S_{\text{sep.}1}$ will become externally manifest as a definite amount of heat or of mechanical work. The same considerations will apply to later stages, in which separation is applied to the fractions $\alpha$ and $\beta$, if these stages, too, are capable of being conducted reversibly.

Similarly in regard to the reversible mixings. In each of their stages the external heat or work effect again varies continuously with the degree of dissimilarity of the components $A$ and $B$. This is so even though the infinite sum $\sum_{i=1}^{\infty} \delta S_{\text{mix.}i} = 2R \ln 2$ is constant and is independent of the degree of dissimilarity of the components. That there is no incompatibility in these statements may be seen by considering the quantity $(2R \ln 2 - \varepsilon)$, where $\varepsilon$ is relatively small. The attainment of this entropy of mixing (falling short of $2R \ln 2$ by not starting the mixing with pure $A$ and $B$) will require a smaller *finite* number of stages if $A$ and $B$ differ considerably than if they differ only slightly. Constancy at the value $2R \ln 2$ only obtains at the unattainable limit of an infinite number of reversible stages. More will be said about this point shortly.

Let us briefly return to the discussion of the first stage in a separation process. The matter can be made much simpler if it is supposed that the stage has been controlled in such a way that the fractions $\alpha$ and $\beta$ are equal in amount, i.e. so that they each contain 1 mole of $A + B$. It then follows from (4.12) that:

$$n_a^\alpha = n_b^\beta; \quad n_a^\beta = n_b^\alpha \tag{4.15}$$

Under these circumstances the number of variables on the right-hand side of (4.14) can be reduced to one. A convenient choice of this variable is:

$$E_1 \equiv (n_a^\alpha - n_b^\alpha)/(n_a^\alpha + n_b^\alpha) \; [= -(n_a^\beta - n_b^\beta)/(n_a^\beta + n_b^\beta)] \tag{4.16}$$

This is zero when $n_a^\alpha = n_b^\alpha$, corresponding to zero separation, and is unity when $n_b^\alpha = 0$, corresponding to the (unrealisable) case of complete separation. Thus $E_1$ is a measure of the separation efficiency in the first stage – and thereby is a measure of the distinguishability of the substances $A$ and $B$ in this stage.

Substituting (4.16), together with the defining equations for the mole fractions, into (4.14) we obtain:

$$\delta S_{\text{sep.1}} = -R\{(1+E_1)\ln(1+E_1)+(1-E_1)\ln(1-E_1)\} \qquad (4.17)$$

Here again it can be seen that there is complete continuity in the value of $\delta S_{\text{sep.1}}$ in relation to the distinguishability of the components $A$ and $B$ which determines the value of $E_1$. As has been said, each of the fractions $\alpha$ and $\beta$ can be submitted to a second, and further, separations. Equations of the type of (4.17) will again be obtained, involving efficiencies $E_2$, $E_3$, etc. Of course, it may no longer be useful to obtain the second, and later, stage fractions as equal in amount and, if so, the efficiencies $E_i$ ($i > 1$) must be defined a little differently from $E_1$. Nevertheless, it will be possible, by carefully designing the separation procedure, to obtain *almost* pure $A$ and $B$ in a sufficient number of stages and after any *finite* number, $n$, of such stages the sum, $\sum_{i=1}^{n} \delta S_{\text{sep.}i}$, will depend on the degree of dissimilarity of the components. It is only at the unattainable limit of an infinite number of stages that this sum reaches the value $-2R\ln 2$ which is independent of the degree of dissimilarity of substances $A$ and $B$. (This follows, of course, from the fact that if the separated material is allowed to re-mix spontaneously, the entropy of re-mixing – equal to the negative of the total entropy of separation – is as given by equations (4.10) and (4.11).)

Even so, the infinite sum $\sum_{i=1}^{\infty} \delta S_{\text{sep.}i}$ is zero if the components are the same ('identical') and are thus inseparable. Thus at the infinite limit the discontinuity of the Paradox may be said to reappear and is reminiscent of a situation in the statistical mechanics of phase change where discontinuity is shown to be perfectly sharp only when $N \to \infty$ where $N$ is the number of molecules in the system.

By way of a brief summary of this resolution of the Gibbs' Paradox, the essential point is that the entropy increase in a spontaneous mixing (e.g. by the removal of partitions between quantities of different gases) is not physically manifest and in that sense is 'unreal'. To obtain manifestation as a definite and maximal amount of external heat (or of mechanical work) either the mixing or a subsequent separation must be carried out reversibly. This requires stagewise operation. The entropy change in each stage of the sequence is a *continuous function* of the degree of dissimilarity of the substances, and so, too, is the sum of the stagewise entropy changes up to any finite number. Thus although equation (4.10) is undoubtedly correct in regard to entropy as a 'theoretical entity', the equation, as applied to a spontaneous mixing, does not correspond to anything which is actually observable. In this sense the Gibbs' discontinuity is an artefact due to a physically unattainable limit. (See also a forthcoming paper with M. L. G. Redhead.)

### 4.5. A quantal resolution of the Paradox as applied to 'states'

An interesting quantal resolution was obtained by von Neumann (1955), Klein (1959) and Landé (1965). It is more restricted than is the foregoing thermodynamic treatment since it refers only to the distinguishability, or otherwise, of the quantum states of particles of *the same kind*. Yourgrau *et al.* (1966) are wrong, we think, in saying that the von Neumann treatment provides a *complete* solution of the Paradox since it does not explain why mixtures of different substances all have the same entropy of mixing, however similar or dissimilar those substances are. Nevertheless we give a brief summary below in order to display the significance of separability in von Neumann's ideas, as well as our own.

Von Neumann was concerned with the consistency of quantum mechanics with thermodynamics. With this in mind he showed that if two states $\phi$ and $\psi$ of a system are orthogonal they are capable, consistently with the Second Law, of being separated. But if $\phi$ and $\psi$ are not orthogonal, complete separation cannot be achieved. He went on to remark that this clarifies Gibbs' Paradox since there is continuous transition between situations where $(\phi, \psi) = 0$[39] to situations where $|(\phi, \psi)| = 1$. In the latter case, he said, 'the states $\phi, \psi$ are identical, and the separation is completely impossible.' The degree of separability is thus continuous over the whole range of the variation of the inner product $(\phi, \psi)$.

Klein extended the theory by considering two equal volumes each containing one mole of the same atoms in states $\phi$ and $\psi$ respectively which correspond to the same energy.[40] These atoms were allowed to mix, at constant total volume, and the entropy increase was shown to be

$$\Delta S/R = 2 \ln 2 - (1 + \alpha) \ln (1 + \alpha) - (1 - \alpha) \ln (1 - \alpha) \qquad (4.18)$$

where $\alpha = (\phi, \psi)$. The entropy of mixing is thus $2R \ln 2$ for the orthogonal case and is zero when $\alpha = 1$, with continuous transition between the two extremes. The result (4.18) is seen to be similar to our own equation (4.17) when due allowance is made for the different definitions of $\alpha$ and $E_1$.[41]

If the distinction between the states $\phi$ and $\psi$ is to be made observable, it is necessary, of course, to use some apparatus whose performance depends on the value of the inner product. Landé offered a nice example: the behaviour of a stream of silver atoms passing through a Stern–Gerlach apparatus. 'Silver atoms pointing north,' he wrote, 'cannot be totally separated from those of spin direction enclosing an angle $\theta$ with respect to the north.' He therefore defined an 'equality fraction' $P = \cos^2(\theta/2)$ and proceeded to show that the entropy of mixing of such atoms, after the merely partial separation which is possible in the Stern–Gerlach apparatus, is given by (4.18) with $\alpha$ replaced by $P$.[42]

Before leaving Landé's example, it may be added that it provides a good instance of the *context dependence* of 'distinguishability' which has already been referred to; the distinguishability of the spin states of the silver atoms depends on the existence of an apparatus which can differentiate between them. Other instances relate to accessibility as discussed in §2.3. Suppose that there are two particles of the same kind in separate boxes each of volume $V$. This implies that the quantum states accessible to the particles are not the same as the joint system would have if the atoms could move freely within the total volume $2V$. The macroscopic distinguishability of the boxes confers a distinguishability on the particles which they do not have when they are together. On the other hand, if the boxes have a wall in common, and if this wall is slightly permeable, the previously inaccessible states would have to be deemed as being accessible over a sufficiently long period of time. Similar considerations apply to the phenomenon of tunnelling through a potential energy barrier or to the effect of high temperature. For instance, two isomers which are readily distinguishable from each other at moderate temperatures may become interconvertible at higher temperatures such that many of the molecules have energies great enough to allow them to pass, at an appreciable rate, over the energy barrier separating the isomers. In short, the time period during which quantum states may, or may not, be accessible to each other is highly relevant to the consideration of distinguishability.[43]

### 4.6. Identity, non-individuality and indistinguishability

The foregoing sections provide a basis for some discussion on the distinctions of meaning which exist between the words occurring in the above heading and on whether or not the usage of those words involves subjective factors.

The reader will recall certain traditional philosophical themes such as the Principle of the Identity of Indiscernibles and the differences between 'particulars' and 'universals'. Most of the philosophical debate has been at the macro-level – the world of tables, chairs, persons and so on – but we shall here give our main attention to the special problems which arise at the atomic level.

The notion of identity is often used in the restricted sense of self-identity, i.e. the identity of a thing with itself as when one speaks of the self-same penny. This usage excludes the notion of replicas – that is of the concept, with which Leibniz seems to have been concerned, of the possible existence of entities identical in all respects except location.[44] Let us leave this aside for the moment.

The main problem about self-identity in the atomic domain is the justification for claiming that some particle remains *the same* particle over a temporal duration. Quinton (1978, p. 66) regarded it as a necessary (although not a sufficient) condition for this relation of sameness to hold, at least at the macro-level, that there exists a continuous spatial path between the locations at which the supposedly self-same entity occurs. As was seen in §4.2, Boltzmann took the satisfaction of this condition as providing the means by which one 'like' particle may be individuated from another, and thus for justifying his combinatorial expression (4.1). However, that expression did not result in a satisfactorily additive entropy. To achieve additivity in the pre-quantum statistical mechanics an *ad hoc* division of the combinatorial formula by $N!$ had to be made. As Gibbs realised, this was tantamount to denying that the atomic particles were capable of being individuated – or at least in the statistical context.

With the advent of quantum mechanics the criterion of the continuity of path was shown to be entirely insecure. For example, when two 'like' particles collide, there occurs an overlap of their wave functions and it becomes impossible to determine which particle was which before and after collision. The overlap also implies a loss of the classical notion of the *impenetrability* of particles; there is no longer a clear sense in which it can be said that the one particle is not in the same element of space as the other.

These matters moved even further away from the classical viewpoint with the development of the 'field version' of quantum mechanics. Entities such as electrons can here no longer be regarded as particles – instead they are states of excitation of one or other of the basic fields and they come and go according to the action of creation and annihilation operators. Thus they do not endure in time – they are not *continuants* – and the notion of self-identity over a duration is inapplicable. The basic particulars are now the points of space–time whose excitations take over the traditional functions of what were previously specks of matter.

This current issue about whether science is best served by an ontology of particles or (as in field theory) by an ontology of events in space–time is a very live one. Obviously enough these alternatives are closely related to relational versus absolute conceptions of space–time respectively. Indeed, what remains unclear in the field theory is how the points of space–time can be individuated if not by referring them, as in the relational view, to axes which are fixed in relation to massive enduring objects such as the Sun and stars.

In this connection some remarks by Strawson (1959) are helpful. It would be a mistake, he wrote, to suppose that the re-identification of locations is something quite different from, and independent of, the re-identification of objects.

> There is, rather, a complex and intricate interplay between the two. For, on the one hand, places are defined only by the relations of things: and, on the other hand, one of the requirements for the identity of a material thing is that its existence, as well as being continuous in time, should be continuous in space.

Before coming to the matter of identity in the other sense – that of the possible existence of exact replicas – let us turn, if only briefly, to the concept of individuality. In fact, it has been used already in the sense of the 'individuation' of entities, even when they are 'of the same kind'.

Individuality, as Post (1963) pointed out, is the concept a person would be using if he were to declare: 'That one there is my umbrella', even though the umbrella in question may be quite indistinguishable from many others and even though that person may be quite mistaken in his claim. In the case of massive entities it is their continuity of existence in space and time which we actually rely on, in claims about individuality, even if this continuity may not be traceable in some actual instance. Yet there also exists the notion of what Post calls 'transcendental individuality' which goes beyond the having of spatio–temporal continuity and indeed goes beyond the having, by the putative individual, of any attributes whatsoever.[45] Does it reside instead in the bare 'substantiality' of physical things, i.e. in that which all material objects supposedly have in common, namely substance or matter? Post rejects this and the conclusion to his discussion, based on the symmetry restrictions and the quantal statistics (§4.7 and §4.8 below), is that fundamental particles cannot be individuated.

This argument creates a problem for those scientists who wish to continue using the notion of particles. It does not make sense, it has been said, to speak of particles unless they can be labelled, even if only conceptually. They must be capable of being *named* as particulars – that is to say, as the *bearers* of properties. (For properties are universals.) If naming is physically impossible, or if the concept of transcendental individuality is too dubious in a philosophical sense, this has been taken by Redhead (1983) as a good argument in favour of the field version of quantum theory, and with it the ontology of events in space–time.

To pursue this matter would be to take it to a deeper level than is appropriate in the present book where the main interest is subjectivity versus objectivity. In what follows we shall continue to write about particles, since this provides such a useful language, even though it has to be recognised that the particles may endure for only limited periods in many instances.

This brings us to the second meaning of identity – that of exact replicas (say, two helium atoms), identical with each other in every respect *except location*. 'Identical with' is clearly a useful concept if only because it is a dyadic relation which is reflexive, symmetric and transitive. As such, it can presumably be used for the purpose of defining equivalence classes and thus for allowing it to be said that certain particles are 'of the same kind' or are 'like' particles. Science could hardly operate without this possibility of classification.

Yet 'identical with' runs into the same difficulties in quantum theory as have already been discussed in connection with self-identity. The loss of the classical notion of impenetrability clearly sets limits to the distinction of entities by location. For this reason one uses instead the concepts of *distinguishability* and of its negation. These are much less problematic and indeed have a relatively straightforward operational character.

Two entities are said to be distinguishable if there is at least one property which can be truly predicated of the one entity but not of the other. The main problem which arises concerns the word 'truly'. For 'indistinguishable from', unlike 'identical with', is not necessarily a transitive relation. For example, two bodies $A$ and $B$ may be indistinguishable in a particular context (such as the measurement of their weight) and similarly $B$ and $C$ may also be indistinguishable in that context, and yet $A$ and $C$ may be capable of being so distinguished. It is a matter of the limit of the objective means of discrimination and this point was prominent in our discussion of Gibbs' Paradox.

In regard to the classification of particles, this matter of non-transitivity gives rise to little trouble in practice so long as the particles in question are fairly stable. Entities, such as electrons, protons, etc., are characterised by a small number of intrinsic properties, such as rest-mass, charge and spin, whose aggregate differs by *finite amounts* from one class of entities to another. For this reason there is little real difficulty in discriminating between the members of different classes. Furthermore, the elementary particles are often *countable*. There are objective techniques, such as the use of scintillation counters, bubble chambers, etc., for achieving this. Millikan's oil drop system provides a good example since the apparatus was able to distinguish between drops charged with a different number of electrons. Thus it can be objectively and verifiably the case that an enclosure contains some definite number of 'like' particles.

As was said in §4.2, Gibbs was the originator of the Indistinguishability Postulate with his remark about the interchange of similar particles not giving rise to a changed state of the overall system.[46] Yet he applied this idea only in a statistical context and he maintained that it remains possible to identify separate particles. This was in the tradition of classical mechanics but is now repudiated, as has been seen above, in quantum mechanics.

Dirac was largely responsible for the more far-reaching modern version of the Postulate. In his book (1958) he put it as follows:

> If a system in atomic physics contains a number of particles of the same kind, e.g. a number of electrons, the particles are absolutely indistinguishable one from another. No observable change is made when two of them are interchanged.

As has been seen, 'particles of the same kind' are just those particles which are indistinguishable from each other. Thus Dirac's first sentence is really a tautology and the physical content of his formulation of the Postulate is to be found in his second sentence. This can be paraphrased to the effect that the interchange of two indistinguishable particles *makes no change* in the expectation value of any observable. Let us put this a little more formally following Redhead (1983):

> In the case of an assembly of indistinguishable particles the expectation value of any observable in any state $|\psi\rangle$ is the same as it is in the state $P|\psi\rangle$ where $P$ is any permutation operator acting on the 'labels' of the particles occurring in the specification of $|\psi\rangle$.

Here it is tacitly supposed that the particles can indeed be labelled, i.e. as if they were endowed with the individuality already referred to, but the Postulate makes the very important assertion that the interchange of the hypothetical labels can make no alteration to any observable property of the assembly. The way in which this physical hypothesis leads on to the symmetry restrictions, and thence to the Bose–Einstein and Fermi–Dirac statistics, will be dealt with in the next two sections.

Perhaps it should be added that indistinguishability is an *open* concept in so far as it leaves undecided whether or not some distinction between entities may later be achieved in some manner which, at an earlier stage, was quite unforeseeable. One can recall the discovery of the existence of isotopes as an instance of a distinction between atoms which was previously unknown. As was seen in Chapter 1, this entailed the introduction of an entropy of isotope mixing – an entropy which can be made physically manifest as heat (i.e. as an 'observable') through a reversible separation of the isotopes.

By way of a brief summary it may be said that the concepts of identity and of indistinguishability do a different sort of work. For example, someone might say: 'The penny which Jones now shows me in his right hand is indistinguishable from the one he showed me a moment ago in his left hand, but it might not be *the same* penny.' Thus, as has been said, indistinguishability has an operational meaning whereas identity refers to what may 'really' be the case, even if this is unprovable.

What is important in the context of this book is that the concept of indistinguishability can be applied without involving human powers of discrimination. Instruments alone are sufficient for the purpose of deciding, at any particular stage of science, whether or not certain entities appear to be 'of the same kind'. The phenomenon of scattering is particularly striking in this respect since it has been shown by experiment, as well as from theory, that like and unlike particles do not scatter after collision in the same way. They obey different laws.[47] For example, the scattering cross-section in the low energy collision of two spinless bosons is four times greater if they are indistinguishable than it would be if they were of different kinds (Messiah, 1964, p. 607). Results such as these support the view that distinguishability and its negation have a fully objective character.

## 4.7. The symmetry restrictions

The following will be concerned only with *non-localised* systems – that is to say, systems consisting of freely moving particles such as exist in gases or as the conduction electrons in metals. If the system contains like particles only, the interchange of any two of them, according to the Indistinguishability Postulate, does not change any observable. *Localised* systems, such as crystals, are different in this respect. For, although the atoms or molecules which occupy the lattice sites are indistinguishable, if they are of the same kind, the sites themselves are distinguishable by virtue of their coordinates relative to the crystal as a whole. This enables the statistical mechanics of crystals to be worked out in terms of the normal modes of vibration of the entire crystal.

For purposes of simplicity the following will also be restricted to perfect gases. The quantum statistics of gases can be developed by either of two methods: (1) An older method in which one first regards the 'like' particles as capable of being 'labelled' and one then reconstructs the wave function, or ket, in such a way as to eliminate that supposition; (2) The method of quantum field theory which makes no reference at its foundations to the existence of particles, whether distinguishable or not.

We here adopt the first method as used by Dirac (1926) and later modified in his book. It follows on naturally from the axiomatic structure of §2.2, based on an ontology of particles, whereas the use of the second method would require a considerable recasting of that structure. To be sure, the possible existence of paraparticles does not fit easily into the scheme we adopt but some mention of these will be made later.

Our objective, as elsewhere, is simply to show that no subjective factors are involved. For this purpose it will be sufficient to discuss a perfect gas consisting of *only two* indistinguishable particles. Like Dirac (1926), we shall use wave function notation (Appendix 2.1) in place of the bras and kets which he subsequently adopted.

Since the particles are mechanically independent of each other we can speak of 'one-particle' states. Let $i$ and $j$ be two of them and let the particles be conceptually labelled 1 and 2. There are then *four* product wave functions for the system as a whole:

$$\text{(a) } \psi_i(1)\psi_j(2) \quad \text{(c) } \psi_i(1)\psi_i(2)$$
$$\text{(b) } \psi_i(2)\psi_j(1) \quad \text{(d) } \psi_j(1)\psi_j(2)$$

and these are orthogonal and span a four-dimensional vector space.

Now it will be seen that (a) and (b) are obtained from each other by a permutation of the supposed labels. According to the Indistinguishability Postulate these two product wave functions give rise to no difference in any observable. It can be argued that the system has only *three* distinct states arising from the pair of one-particle states $i$ and $j$.

However, there is a physically more useful way of looking at the matter. The Superposition Principle allows of any desired linear combinations of (a) and (b) and such combinations represent permissible vectors just as well as do (a) and (b) themselves. Dirac examined the following two superpositions:

(e) $\psi_{ij}^s = \{\psi_i(1)\psi_j(2) + \psi_i(2)\psi_j(1)\}/2^{1/2}$

(f) $\psi_{ij}^a = \{\psi_i(1)\psi_j(2) - \psi_i(2)\psi_j(1)\}/2^{1/2}$

where the superscripts $s$ and $a$ denote 'symmetric' and 'anti-symmetric' respectively and the division by $2^{1/2}$ is to achieve normalisation. It will be seen that the interchange of labels (1) and (2) changes the sign of $\psi^a$ but does not change the sign of $\psi^s$. Does this mean, contrary to what was said above, that the *given pair* of particles can really enjoy *four* states, i.e. (c), (d), (e) and (f)?

That this would be a false conclusion follows from the fact that a mere change of sign of a wave function has no effect on the value of any observable since these values depend on the *squares* of wave functions (§2.2). Could it be the case, however, that symmetric wave functions, i.e. those such as (c), (d) and (e) which do not change sign when the particle 'labels' are interchanged, correspond to one class of particles and that anti-symmetric wave functions such as (f), which do change sign, correspond to an entirely different class?

That this is a possibility may be seen as follows. The Indistinguishability Postulate implies that the system's energy, and hence also its Hamiltonian, must be unchanged, like any other observable, if the particle labels are permuted. Now the temporal development of the system, as given by the time-dependent Schrödinger equation (Appendix 2.1), depends on the Hamiltonian together with $\psi$, and not the square of $\psi$. This means that if the wave function for the two-particle system has any of the symmetric forms (c), (d) and (e) the system *always* has a wave function of this form. Similarly, if the system has the anti-symmetric form, (f), the system remains permanently in that form. In other words, the natural world, if quantum mechanics is correct, is such that the symmetric and the anti-symmetric wave functions are entirely inaccessible from each other.

The conjecture that there exist two classes of particles in regard to the symmetry or anti-symmetry of the wave function has been amply corroborated by experiment, especially by the study of the spectra of molecules. If the symmetric and anti-symmetric states were not mutually inaccessible for a particular molecule, as is required by the Indistinguishability Postulate, theory would predict many more spectral lines than are actually observed. The 'forbidden' transitions do not occur. Further experimental corroboration has been obtained from the thermodynamic behaviour of photons and of the conduction electrons in metals. This will be discussed in §4.8.

The classes of the *bosons* and the *fermions* are those classes of particles which are restricted to symmetric and anti-symmetric states respectively, and the distinction is known to be connected with spin. Electrons, positrons, protons, neutrons and other nucleons having half integral spins are fermions whereas entities such as photons and pions having zero or integral values of their spins are bosons. The boson or fermion character of compound particles, such as atoms and molecules, can be deduced (so long as they interact only weakly) from their overall spin, as determined by the elementary particles of which they are made up. For example $He^4$, $H_2$ and $D_2$ which contain an even number of electrons, protons and neutrons, are bosons. Those such as $He^3$ and HD which contain an odd number are fermions.

A further consequence of the Indistinguishability Postulate, which again supports its fully objective character, is the Pauli Exclusion Principle. Considering the two-particle states (a) to (f) it will be seen that, if $i=j$, $\psi^s$ becomes proportional to (c) and (d) whereas $\psi^a$ becomes *zero*. It follows that no physical reality can be attributed to a situation in which two fermions within an enclosure are in the same eigenstate. This restriction, which applies only to fermions, has been amply corroborated in chemistry, notably in regard to the structure of the elements.

It is unnecessary for present purposes to go into the construction of $\psi^s$ and $\psi^a$ for the case where there are more than two 'like' particles in the enclosure. A more interesting issue relates to necessary and sufficient conditions. As has been seen, wave functions of the type (c) to (f) provide sufficient conditions for the satisfying of the Indistinguishability Postulate. But do they also provide the necessary conditions? Some uncertainty over this issue was due to a confusion about whether the Indistinguishability Postulate should be taken as being concerned with *quantum states* or with *observables*. Dirac had early expressed the view that the latter is the case and, indeed, he had envisaged the possibility (1958, pp. 209–11) that, when the number of particles is greater than two, there may be vectors, other than the fully symmetric and anti-symmetric vectors, which result in unchanged values of the observables when 'labels' are permuted.

A number of scientists[48] subsequently examined this possibility in much greater detail and the results of these studies suggested the possible existence of classes of particles – the so-called paraparticles – which are neither bosons nor fermions. So far, no such particles have been found for certain, and, indeed, it is known that all of the relatively familiar and stable particles fall into the two original classes. If this turns out to be a general conclusion it would appear that some restrictive condition will be needed within the structure of quantum mechanics to ensure that the fully symmetric and anti-symmetric wave functions are both necessary and sufficient for the satisfying of the Indistinguishability Postulate.

## 4.8. Bose–Einstein (BE) and Fermi–Dirac (FD) statistics

The previous section leads on to the question, more relevant to this book, about how the symmetry restriction affects the statistics of systems consisting of 'like' particles. The mutual inaccessibility of symmetric and anti-symmetric states results in two kinds of statistics, one for bosons and the other for fermions.

As in §4.7, the matter will first be illustrated by reference to a system containing only two such particles and again with the proviso that they interact sufficiently weakly for the wave function of the compound system to be expressible as the product of 'one-particle' wave functions – a condition which is satisfied by dilute gases. Attention will be confined once more to only two eigenstates, $i$ and $j$, and these will now be supposed to belong to the same energy eigenvalue.

If the Indistinguishability Postulate did not hold, all four of the products, (c), (d), (e) and (f), would be possible wave functions for the composite system. It would then be concluded, by adopting our usual assumption of equal *a priori* probabilities, that each of the four states has a probability of $\frac{1}{4}$. This corresponds to the 'uncorrected' Boltzmann counting referred to in §4.2.

As soon as the Indistinguishability Postulate is allowed for, a system of two bosons becomes restricted to the states (c), (d) and (e). It follows that the transition from the uncorrected Boltzmann counting to the quantum counting for bosons, due to Bose and Einstein who first applied it to photons, raises the probability of the allowed states from $\frac{1}{4}$ to $\frac{1}{3}$.

In the case of fermions, only (f) is a possible wave function, since this is the only one which changes sign, i.e. is anti-symmetric, on permutation of 'labels'. Its probability is thereby raised to unity and this is characteristic of the Fermi–Dirac counting.

Before leaving this illustrative example, it is of interest to notice the situation which occurs when both particles happen to be 'crowded together' into the same one-particle state, either $i$ or $j$. As has been seen, this cannot obtain when the particles are fermions, but if they are bosons it occurs with *greater* probability than could be accounted for on a classical basis. It might appear as if there were a repulsion between fermions and an attraction between bosons. Einstein expressed his surprise when he appreciated this result[49] and only later was it realised, as a consequence of the formulation of the Indistinguishability Postulate, that particles may not be statistically independent even when, as in a perfect gas, they are mechanically independent.

We turn now to the general treatment of a perfect gas consisting of 'like' particles, $N$ in number. The system will be assumed to have a discrete energy spectrum as is normally the case for an enclosed gas. Although the easiest method for deriving the Bose–Einstein and Fermi–Dirac statistics is by use of the grand partition function, we shall adopt the combinatorial method instead. Our reasons for doing so are that the combinatorial formulae are needed in Appendix 4.2, but also because the method in question shows more clearly the essential difference between Boltzmann counting, on the one hand, and that which is used in the quantum statistics on the other.

This difference can be expressed as follows. In Boltzmann counting, the *molecules* are supposed to be distinguishable and one calculates the number of ways in which they can be distributed among energy 'cells', resulting in equations (4.1) or (4.2). In Bose–Einstein and Fermi–Dirac counting, on the other hand, the 'like' molecules are taken as indistinguishable and one counts the number of observationally distinguishable *states* which are occupied by $0, 1, 2, \ldots$ of the molecules. This is what is required by the Indistinguishability Postulate. The distinction between Bose–Einstein and Fermi–Dirac counting is simply that in the latter the occupancy of each state is limited to 0 or 1 in accordance with the Exclusion Principle which is also a consequence of the Indistinguishability Postulate.

Following Tolman (1938, p. 362), let the energy spectrum of any one of the $N$ 'like' particles be divided into ranges $\Delta \varepsilon_\kappa$ where each range corresponds to the approximate accuracy of specification. Let there be $g_\kappa$ energy eigenstates in the range $\Delta \varepsilon_\kappa$ and let $n_\kappa$ be the number of particles momentarily occupying this group of states.

Taking first the case of bosons where $n_\kappa$ can be any number, we obtain for the number of occupied states for the system as a whole, arising from particular values of $n_\kappa$ and $g_\kappa$, the expression:

$$\frac{(n_\kappa + g_\kappa - 1)!}{n_\kappa! \, (g_\kappa - 1)!} \tag{4.19}$$

this being the number of ways in which $n_\kappa$ indistinguishable objects can be assigned to $g_\kappa$ distinguishable boxes with no restriction on the number per box.[50] The total number of states corresponding to a condition of the system specified by the $n_\kappa$ is the product of (4.19) over all values of $\kappa$:

$$G_{BE} = \prod_\kappa \frac{(n_\kappa + g_\kappa - 1)!}{n_\kappa! \, (g_\kappa - 1)!} \tag{4.20}$$

In the case of fermions the expression comparable with (4.19) is

$$\frac{g_\kappa!}{n_\kappa!(g_\kappa-n_\kappa)!} \tag{4.21}$$

which is the number of ways in which $n_\kappa$ indistinguishable objects can be assigned to $g_\kappa$ distinguishable boxes with no more than one object per box. Thus in place of (4.20) we have

$$G_{\mathrm{FD}}=\prod_\kappa\frac{g_\kappa!}{n_\kappa!(g_\kappa-n_\kappa)!} \tag{4.22}$$

By contrast, the expressions for the 'uncorrected' Boltzmann counting are obtained from (4.2) and in the present notation are

$$\frac{N!}{n_1!\dots n_\kappa!\dots}\cdot g_1^{n_1}\dots g_\kappa^{n_\kappa}\dots \tag{4.23}$$

and

$$G_{\mathrm{MB}}=\prod_\kappa\frac{N!\,g_\kappa^{n_\kappa}}{n_\kappa!} \tag{4.24}$$

By using the method of 'the most probable distribution', together with the use of undetermined multipliers and Stirling's approximation, it can then be shown that in the equilibrium state the $n_\kappa$ are given by:

$$n_\kappa=\frac{g_\kappa}{\exp\left(\alpha+\beta\varepsilon_\kappa\right)+C} \tag{4.25}$$

where $\beta=1/kT$ and $C$ takes the values $-1$, $+1$ and $0$ for Bose–Einstein, Fermi–Dirac and Maxwell–Boltzmann statistics respectively. $\alpha$ is a constant and is zero in the case of photons, since these 'particles' obey no conservation law, $\sum_\kappa n_\kappa=N$, and correspondingly there is no undetermined multiplier.

Little more needs to be said since it has been seen that the quantum statistics introduces no subjective factors not already discussed in Chapter 2. Let us add only a few remarks about the extent to which the Bose–Einstein and Fermi–Dirac statistics have been confirmed by experiment.

Earlier in this section we discussed an example with $g_\kappa=2$ and $n=n_\kappa=2$ which was specially chosen to display a certain point.[51] However, that example was not very realistic and under the circumstances normally prevailing in *molecular* gases, $g_\kappa\gg n_\kappa$. When this is so, it can be shown that (4.25) gives almost exactly the same results whichever value of $C$ is appropriate except at temperatures close to the absolute zero. The three forms of statistics thus do not differ significantly for molecular gases under normal conditions. In particular, the Sackur–Tetrode equation (4.5) remains valid in Bose–Einstein and Fermi–Dirac statistics except that a numerically small correction is needed if the spin states are degenerate.

On the other hand, in the case of 'gases' whose particles are much smaller in mass than molecules the various statistics differ very appreciably. One important example is that of the conduction electrons in metals. These have been found to behave as a fermion gas in good agreement with Fermi–Dirac statistics. A more important example from the historical viewpoint was that of radiation regarded as a boson gas of photons. This, of course, was the origin of the Bose–Einstein statistics. By putting $\varepsilon_\kappa = h v_\kappa$ together with $\alpha = 0$, equation (4.25) leads almost immediately to the Planck black body radiation law which has been well corroborated by experiment. It is these two examples which provide the most reliable support for the quantum statistics.

As has been said already, the Indistinguishability Postulate itself has obtained much wider confirmation than is provided by statistical mechanics; notably from the spectra of a great variety of molecules and from scattering processes.

## 4.9. From indistinguishability to distinguishability

It remains to answer a question raised in §4.1: Is there a continuous transition in regard to indistinguishability in passing from the microscopic to the macroscopic domains? It was natural in classical mechanics for *all* bodies to be regarded as individuals, if only because large ones can certainly be followed in their trajectories. There seemed no good reason why very small bodies should not also be individuals. Thus classical mechanics led to the problem of how the apparently necessary assumption of indistinguishability at the atomic level was to be explained.

The situation was reversed in quantum mechanics. For, whereas classical theory sought to understand the micro-level in terms of the laws derived from the macro-level, quantum mechanics proceeds in the reverse direction. The indistinguishability of, say, two electrons was taken as being almost self-evident and it led to a host of satisfactory predictions. Yet it still left a problem, the reverse of the classical one. This is to explain the distinguishability of massive entities when it is assumed that they are made up of indistinguishable elementary particles.

We first put forward a qualitative argument to the effect that indistinguishability goes over continuously into distinguishability, and that it does so in a fully objective manner which is independent of human powers of discrimination. This is followed in Appendix 4.2 by a mathematical treatment.

For these purposes we shall no longer be concerned with *gases* and this is because one macroscopic sample of a gas is not, in fact, distinguishable from another macroscopic sample of the same gas. What are distinguishable are *the vessels* in which those samples are contained. In other words, when it is asserted that macroscopic bodies are distinguishable there is a restriction to bodies which are capable of retaining their own spatial boundaries, at least temporarily. Thus it applies to vessels, billiard balls and stars, and even to clouds, but it does not apply to gases.

Let us consider the building up of bodies of this kind from the atomic particles of which they are composed. A suggestive starting point is the building up of large molecules. Consider, for example, the saturated hydrocarbons (the alkanes) whose general formula is $C_nH_{2n+2}$ and whose first and second members are methane, $CH_4$, and ethane, $C_2H_6$, respectively. Already, at the fourth member, butane, there are two alternative structures:

$$H_3C-\underset{\underset{H}{|}}{\overset{\overset{H}{|}}{C}}-\underset{\underset{H}{|}}{\overset{\overset{H}{|}}{C}}-CH_3 \quad \text{and} \quad H_3C-\underset{\underset{CH_3}{\diagdown}}{\overset{\overset{CH_3}{\diagup}}{C}}-H$$

The next member has three such isomers and the one after that has five. Thereafter the number of structures for a given number of carbon and hydrogen atoms rises very rapidly. For example, there are 802 isomers of $C_{13}H_{28}$ as calculated by Cayley.

The point we are making is this: even though the constituent atoms are indistinguishable, the putting together of large molecules gives rise to a rapidly increasing variety of structures which *are* distinguishable, both chemically and physically. Think, for example, of two balls of paraffin wax which are so perfectly alike to the eye that they might appear as replicas of each other. Yet, in fact, they are almost certainly very different – and verifiably so. Since paraffin wax is a mixture of many alkanes, particularly from $C_{22}H_{46}$ up to $C_{28}H_{58}$, each having an immense number of isomers, it is exceedingly improbable that the chemical constitution of the two balls is the same.

Such considerations apply *a fortiori* to the very large biological molecules such as DNA. The immensity of the number of distinct DNA structures is one of the factors which ensure that each living creature is almost certainly genetically different from every other member of its own species.

Yet it might be objected that these examples of chemical isomers are a rather special case. Let us think instead of the building up of a ball of, say, copper from its single atoms. Even if the ball could be regarded as being a single crystal there would still remain an immense scope for differences between one such ball and another, due to the unequal distribution within them of lattice defects. And, in fact, such balls are much more likely to have a mosaic structure of minute crystals. This factor again gives rise to an immense number of structural differences between any two balls, differences which are detectable by use of microscopy.

The conclusion to be drawn is that there can be no real identity between any two apparent replicas of a macroscopic body. A sufficiently detailed examination would reveal that the apparent replicas are capable of being distinguished – not simply by location but also by their internal structures.

Putting the foregoing in quantum terms, it may be said that, with increasing size and complexity of a body, the number of its possible quantum states increases at an immense rate. The probability that any two seeming replicas of the body will be 'in' the same quantum state thus becomes vanishingly small.

Consider two freshly minted pennies, 1 and 2. Even though they may be indistinguishable to the eye they are 'individuals' by virtue of their differing internal structures, as well as by the fact that they have distinguishable locations and trajectories. If they were tossed repeatedly, it would be expected that two heads (HH) and two tails (TT) would each be obtained on about $\frac{1}{4}$ of the occasions and that a head and a tail (HT) would be obtained on the remaining $\frac{1}{2}$ of the occasions – the latter arising from the fact that $H_1T_2$ and $H_2T_1$ each have a probability of $\frac{1}{4}$ because of the distinctness of the pennies. On the other hand, if the coins could be treated as being *indistinguishable* bosons the probabilities of HH, TT and TH might be expected to be $\frac{1}{3}$ each, and if they could be treated as indistinguishable fermions the probabilities of HH and of TT might each be expected to be zero, whilst that of TH has increased to 1. This is to extend to coins the same probabilistic considerations as were applied in §4.8 to two 'like' particles having states $i$ and $j$.

This idea of applying Bose–Einstein and Fermi–Dirac statistics to the pennies clearly leads to conclusions which are contrary to experience, and this is due to the fact that the pennies are distinguishable entities to which the uncorrected Maxwell–Boltzmann statistics should be applied. The question arises whether we can arrive at a *continuous* transition from the Bose–Einstein and Fermi–Dirac statistics, which are the true statistics of elementary particles, to the uncorrected Maxwell–Boltzmann statistics which apply to macroscopic entities.

The foregoing considerations suggest that the desired continuity is to be looked for in a gradual and continuous increase in the number of degrees of freedom which are associated with the internal structure of the body in question. If so, the consequence would be that macroscopic observables, of which the 'heads up' and 'tails up' of coins is but one instance, become enormously degenerate. That this is so is demonstrated in Appendix 4.2, where it is shown that the probabilities of particular values of observables vary continuously, as bodies are made larger and larger, and change smoothly from the values predicted by the Bose–Einstein and Fermi–Dirac statistics to those predicted by the original Maxwell–Boltzmann counting.[52]

Let us conclude that out of the terms 'identity', 'non-individuality' and 'indistinguishability', the third is the most useful in physical science.[53] It plays a significant part in the solving of two of the early problems of statistical mechanics – namely Gibbs' Paradox and the obtaining of an additive entropy – and it is also of great value in regard to the classification of elementary particles. In the form of the Indistinguishability Postulate, which appears as being physically although not logically necessary, it figures preominently in the quantum mechanics of the present time. In all of these applications the notion of indistinguishability has been found to be fully objective.

# 5

# *Entropy and information theory*

### 5.1. Introduction

In 1949 Claude Shannon put forward an important measure of the 'amount of information' which is contained in messages sent along a transmission line. This measure, denoted $H$, was subsequently applied in fields remote from communications engineering, and indeed is now regarded as a useful characteristic of any probability distribution $\{p_i\}$. Jaynes later proposed that the maximisation of $H$, subject to any constraints dictated by the particular application, is the best means for determining the prior probabilities, $p_i$, in a least biassed manner. This idea has been advocated as being a general principle of statistical inference and has been used in economics, engineering and almost all fields where statistical methods are required.

Shannon's measure in conjunction with Jaynes' Principle can also be used for the purpose of providing a mathematical formalism for statistical mechanics, and this has been done by Jaynes himself and by Tribus, Katz, Hobson and others. The consequence has been that an earlier subjective view of entropy, traceable to Maxwell and already discussed from various angles in §§ 1.3, 2.3, 2.5 and 3.8, has once again gained ground. Since it has been the main purpose of this book to refute that view, it behoves us in this final chapter to examine the impact of information theory on thermo-dynamics and statistical mechanics during the last two decades. To be sure, this will not be a matter of criticising the mathematical formalism but only of examining the interpretations which have been placed upon it.

### 5.2. Shannon's measure

Consider a 'message' consisting of no more than four repetitions of either 0 or 1. Possible messages are 0000, 0001, etc., and there are $2^4 = 16$ of them. As soon as any one of them has been singled out there is no longer any uncertainty as to what the message is. Provided that the 0s and 1s are equally probable it is convenient to define the 'information' in the chosen message as being $\log_2 2^4 = 4$, so that the information per symbol is unity.

There are good reasons for choosing a logarithmic measure and for adopting logarithms to the base 2. In the first place, it seems reasonable to maintain that, say, three transmission channels should be capable of handling three times as much information in a given time as a single channel. For example, the triple channel should be capable of dealing with a compound message of 12 symbols in the time in which a single channel deals with only four. Now there are $2^{12} = 16^3$ different 12-symbol messages, made up from 0s and 1s, and thus the information contained in any one of them, according to the above definition, is $\log_2 16^3 = 12 \log_2 2 = 12$. This, as is desired, is just three times as much as the single channel can deal with.

Going back to the case of the single channel, it will be seen that the 'choice' of one particular message out of 16 possibilities involves four binary decisions – one for the choice of a 0 or a 1 in the first position in the message, a second decision for the choice of the symbol in the second position, and so on. The use of a logarithm to the base 2 thus ensures that each binary decision contributes unity to the 'information'. It is said that the four-symbol message has information amounting to four 'bits', where the latter word is short for 'binary digits'. It may be noted at this point that the notion of 'choice' or of 'decision' need not be taken in a subjective sense; an automaton could choose the message just as well as a person.

Consider a message which is $N$ symbols in length and is chosen from the letters $A, B, C$ and $D$. Since there are $4^N$ possible messages, any one of them will carry the information $N \log_2 4 = 2N$. The information *per symbol* is now 2. However, in the transmission of any *real* message it will be necessary to allow for such differences as may exist in the intrinsic probabilities of the various symbols. For example, if the message is in the English language the letter $E$ will occur far more frequently than the letters $Q, X$ or $Z$.

It was to provide for this situation that Shannon (1949) elaborated the simple logarithmic formula and proposed the following function, $H$, as a suitable measure for the information per symbol:

$$H = - \sum_{i=1}^{n} p_i \log_2 p_i \qquad (5.1)$$

Here $n$ is the number of *kinds* of symbols which might be employed in the message and $p_i$ is the probability of the $i$th kind. For example, if $A, B, C$ and $D$ are the only available symbols and have probabilities $\frac{1}{2}, \frac{1}{4}, \frac{1}{8}$ and $\frac{1}{8}$ respectively, then $H = \frac{7}{4}$. This is lower than the previously calculated value of 2 when all four symbols were taken as being equally probable. It is quite generally the case that $H$ has its largest value when the symbols have equal probabilities, and (5.1) then reduces to the simple logarithmic formula. It will be recalled that $S_G$ in equation (2.15) similarly reduces to $S_{BP} = k \ln W$ under the same circumstances.

Further elaboration of (5.1) is needed to allow for the influence of one symbol on another, e.g. to allow for the changed probability of $U$ in English when $Q$ is the immediately preceding symbol. Conditional probabilities must then be used in (5.1), but this is not a matter which need be gone into for present purposes.

It was shown by Shannon himself, and also by Khinchin (1957), Feinstein (1958) and Hobson (1971), that the $H$ function is a unique and self-consistent measure in the sense that it is the *only* function which satisfies certain conditions which would be reasonably expected of a measure of information. Khinchin's choice of conditions will be quoted here since these have an important bearing on a point to be raised later. (Khinchin's third condition is perhaps a little less perspicuous than an equivalent one as used originally by Shannon.) Referring to 'events' rather than to 'symbols' the conditions are as follows:

1. For given $n$ and for $\sum_{i=1}^{n} p_i = 1$, the required function $H(p_1, \ldots p_n)$ shall take its largest value when all $p_i$ are equal to $1/n$.
2. $H(p_1, \ldots p_n, 0) = H(p_1, \ldots p_n)$. That is to say, the inclusion of an impossible event, or any number of them, makes no alteration to the value of $H$.
3. If $A$ and $B$ are two finite sets of events, not necessarily independent, the value of $H$ for the occurrence of joint events shall be the value for the set $A$ alone plus the expectation value of the additional information given by the $B$-type events after the occurrence of $A$-type events. More precisely

$$H(AB) = H(A) + \sum_i p_i H_i(B)$$

where $H_i(B)$ is the conditional value of $H$ for $B$-type events given that $A_i$ has occurred with probability $p_i$.

As has been said, it can be proved that (5.1) is the only function which satisfies these, or equivalent, conditions.

Now Shannon was careful to point out that he used the term 'information' in a highly technical sense and that it bore no necessary relation to *meaning*. For example, the message 'COMING' has no greater value of $H$ than has some alternative, and quite meaningless, arrangement of the same symbols. Apparently his own inclination was to name his function $H$ either as 'uncertainty' or as 'information', both terms being taken in a technical sense. *H* is a measure of the uncertainty of the message *before* it is chosen, out of all the alternatives from the same set of events; equally it is a measure of the information *after* the message has been chosen.

Yet he was subsequently persuaded by von Neumann to call it entropy! 'It is already in use under that name,' von Neumann is reported to have said (Tribus, 1963), 'and besides it will give you a great edge in debates because nobody really knows what entropy is anyway.'

Thus confusion entered in and von Neumann had done science a disservice! To be sure, there are good mathematical reasons why information theory and statistical mechanics both require functions having the same structure. Yet this formal similarity of *H* and $S_G$ does not signify that the two functions necessarily refer to the *same situation*. They may do so in some contexts but not in others. As Popper (1974) very clearly put it, it is a matter of whether the $p_i$ 'are probabilities of the same attributes of the same system'.

Irrespective of which of these names is used to denote the *H* function, certain difficult questions suggest themselves. For instance: *Whose* information or uncertainty is being referred to? *Where* does the entropy reside? Few of the information theorists have dealt with such questions at all adequately, although Jaynes (1979) admits their cogency.

Suppose Jones sends the message 'COMING'. Obviously he knew the message he wanted to send and didn't select it from among the $26^6$ alternative six-letter combinations available from the alphabet. So the value of *H* as calculated from Shannon's formula is not a measure of Jones' uncertainty or information. But neither is it a measure of the recipient's uncertainty or information. He may have expected a message from Jones, but not one which would be *any* jumble of symbols – only one which would convey meaning. Perhaps he expected only COMING or NOT COMING. If so, the message *for him* carried *only one bit* of information, and not the much larger number of bits which would be calculated from Shannon's formula. The problem about whose information is measured by *H* remains obscure.[54]

A similar difficulty arises if the term 'entropy' is used. When we speak of thermodynamic entropy we localise it (even though it is not conserved) in some physico–chemical system. It can also be made physically manifest as a maximal heat intake by that system. No such possibility of location, or of a 1:1 relationship with heat, is available in the kind of situation to which Shannon originally applied his measure, i.e. to transmission lines, or, indeed, in most other situations to which information theory is applied. In short, we are clearly no longer dealing with entropy in its original thermodynamic sense.

These problems are surely of very considerable significance. Difficulties arising from the choice of a name for a concept indicate a basic ambiguity about what that concept really means, or about whether it is being stretched too far. For example, unlike Shannon's situation, we are not concerned in statistical mechanics with 'outcomes'; by specifying the thermodynamic variables of a system we do not 'choose' its quantum state! Thus in the thermodynamic context $H$ (or $S_G$) is not a measure of the information contained in an outcome. To be sure, it could be regarded as a measure of the 'uncertainty' of the actual quantum state. Yet this would be to give an entirely false emphasis if it were taken as a measure of *our* ignorance. It is irrelevant to bring 'us' into the picture since, as was argued in Chapter 2, the $p_i$ of statistical mechanics are *objective* probabilities.

It would, therefore, have been far better if the $H$-measure had been given an entirely neutral name. Tisza (1966) suggested that it should have been called 'the dispersal' of the particular probability distribution. This would have been admirable and would have resulted in the avoidance of much confusion. Unfortunately the name 'entropy' for Shannon's measure has become so widespread that it seems hopeless to try to put the clock back. Even so, one must continue to make a careful distinction between the informational entropy, on the one hand, and the originally defined thermodynamic entropy, on the other. Only in certain physical contexts do they become synonymous.

### 5.3. Jaynes' Principle and statistical mechanics

The usual route into statistical mechanics, as described in Chapter 2, is to adopt the conceptual structure provided by quantum mechanics and then to make the postulate of equal *a priori* probabilities and random phases. These refer to the 'events' which are the occurrence of the energy eigenstates of a substantially isolated system at equilibrium and having fixed volume and composition. From that postulate the canonical distribution, $p_i \propto \exp(-\beta\varepsilon_i)$ can be deduced for the case where the system, instead of being isolated, is in thermal equilibrium with a heat reservoir. The quantity $S_G = -k \sum p_i \ln p_i$ is then shown to be a satisfactory entropy analogue and has a maximum value for the canonical distribution.

Starting in 1957, Jaynes began to make an *inverse use* of statistical mechanics, based on its well-established success. Consider any set $\{E_i\}$ of events which are discrete and which are mutually exclusive and jointly exhaustive. Let the set $\{p_i\}$ be their unknown prior probabilities. Jaynes proposed that a very general technique for discovering the *least biassed* distribution of the $p_i$ consists in the maximisation of the Shannon $H$-function, subject to whatever constraints on the $p_i$ are appropriate to the particular situation. The maximisation of $H$ was thus put forward as a *general principle of statistical inference* – one which could be applied to a wide variety of problems such as occur in economics, engineering and many other fields.

This principle was expressed by Jaynes (1957) as follows where the word 'entropy' refers to the $H$-function:

> ... in making inferences on the basis of partial information we must use that probability distribution which has maximum entropy subject to whatever is known. This is the only unbiased assignment we can make; to use any other would amount to arbitrary assumption of information which by hypothesis we do not have.

The matter has been the subject of considerable discussion in the literature. References are given by Cyranski (1978) and also in Jaynes' reply (1979) to his critics. In a recent paper (1981) he has greatly clarified, and somewhat restricted, the kind of problems to which he believes the principle is applicable. Other important papers are those of Band and Park (1976, 1977) where they examine the application of the $H$-function to quantum phenomena. However, these controversies are not our present concern. What is significant in the context of this book is the bearing of Jaynes' papers on the subjectivity/objectivity issue in thermodynamics and statistical mechanics.

As was said in §5.1, Jaynes' Principle clearly provides an alternative route into statistical mechanics itself. For example, let $H$ be maximised, subject only to the trivial constraint $\sum_i p_i = 1$. An application of the Lagrange undetermined multipliers leads immediately to the result that all of the $p_i$ are equal in the least-biassed distribution. The maximisation of the 'informational entropy' may thus be said, although rather disingenuously, to result in a *deduction* of equal *a priori* probabilities, whereas this was a *postulate* in the system of statistical mechanics described in Chapter 2.

However, it would indeed be disingenuous to suppose that the need for a postulate has been avoided by this procedure. For it will be seen from the very first item in Khinchin's list of conditions that the $H$-function has been constructed in just such a way that, after its maximisation, it leads *necessarily* to the same result as was embodied in our basic postulate of Chapter 2. Similar remarks may be made about the use of Jaynes' Principle for the purpose of deducing the canonical distribution where the additional constraint $\langle U \rangle = \sum_i p_i \varepsilon_i$ is applied. The Lagrange multiplier technique leads at once to $p_i \propto \exp(-\beta\varepsilon)$, but again only because the *H*-function implies equal *a priori* probabilities in the absence of this constraint. It may be added that the foregoing also follows from conditions for the uniqueness theorem different from Khinchin's choice, since his first condition follows logically from the alternative sets of conditions.

To summarise, it may be said that if we define

$$H = -\sum p_i \log_2 p_i$$
$$S_G = -k \sum p_i \ln p_i$$

then *H* and $S_G$ are *identical* (after allowing for Boltzmann's constant and the conversion factor between the logarithms) under circumstances where the $p_i$ refer to *the same* probabilities, e.g. to the probabilities of a system's energy eigenstates. What is done in the application of information theory as a route into statistical mechanics is therefore simply to replace the postulate of equal *a priori* probabilities, in the microcanonical distribution, by Jaynes' Principle in conjunction with the postulates of the uniqueness theorem.

Yet there remains a big difference between the kinds of *interpretive language* which are used in the two methods of approach. In Chapter 2 we defended the view that the $p_i$ used in statistical mechanics are fully objective probabilities. However, when probability theory is used in a broader context, the subjective interpretation, as developed by Keynes and Jeffreys, is undoubtedly more comprehensive. Jaynes, wishing to give his principle of the maximisation of *H* the widest possible application, therefore adopted a prevailingly subjective standpoint. For example, as already quoted in Chapter 2, he remarked that the statistical mechanical entropy can be said in a subjective sense to measure 'our *degree of ignorance* as to the true unknown microstate, . . .'.

Two further quotations are of interest. From his first paper: 'The only place where subjective statistical mechanics makes contact with the laws of physics is in the enumeration of the different possible, mutually exclusive states in which the system might be.' And in his second paper of 1957 he said that the term 'irreversible process' is a semantic confusion: 'it is not the physical process that is irreversible, but rather our ability to follow it.'

Jaynes' position has been clarified and modified in his (1979) paper, where he recounts the difficulties he had experienced in explaining his earlier ideas to Uhlenbeck. Apparently Uhlenbeck had protested, as we would have done, that entropy is an objective physical quantity. Only much later did it occur to Jaynes, as he candidly remarks, that his reply to Uhlenbeck should have been as follows:

Certainly, different people have different amounts of ignorance. The entropy of a thermodynamic system is a measure of the degree of ignorance of a person whose sole knowledge about its microstate consists of the values of the macroscopic quantities $X_i$ which define its thermodynamic state. This is a completely 'objective' quantity, in the sense that it is a function only of the $X_i$, and does not depend on anybody's personality. There is then no reason why it cannot be measured in the laboratory.

Needless to say, this is close to our own view. We would go further and assert that the existence of entropy is independent of man's presence in the world – it is objective$_2$ – although any numerical value is naturally subject to human controversy and revision.

### 5.4. Brillouin's Negentropy 'Principle'

An example of how the concepts of information theory can sometimes lead to misleading, or even to positively erroneous, conclusions is provided by the Negentropy 'Principle' which Brillouin also called the 'Generalized Second Law'. This asserts that thermodynamic entropy is, in some sense, interconvertible with 'information', and in particular that a gain of information can be used to achieve a real reduction of entropy and not merely a correction to an assigned entropy value such as was discussed in §1.3. Let us see how Brillouin expressed the matter in chapters 12 and 13 of his book (1962) and in his research papers (1953, 1954).

Using the entropy analogue $S = k \ln W$ he first considers a reduction in a body's entropy (by transfer of heat to another) so that $W$ is made to diminish from an initial value $W_0$ to a smaller value $W_1$. He then defines the quantity $I$:

$$I_1 \equiv S_0 - S_1 = k \ln (W_0/W_1)$$

This is called the 'bound information' about the system in question. If the latter is now left to evolve spontaneously under conditions of isolation, Brillouin writes $\Delta S_1 \geqslant 0$, and hence for the two processes together

$$\Delta(S_0 - I_1) \geqslant 0$$

Defining 'negentropy' by $N \equiv -S$ the inequality is re-written as

$$\Delta(N_0 + I_1) \leqslant 0$$

In words he says: 'The sum of negentropy and information may remain constant in a reversible transformation, and will decrease otherwise.'

Little exception can be taken to this procedure so long as the term 'bound information' is regarded as nothing more than *a name* given by Brillouin to the entropy change $(S_0 - S_1)$, a decrease which he admits 'must be furnished by some external agent whose entropy will increase'. However, something much more significant than a mere naming is involved when he proceeds to write the 'reversible reaction $I \rightleftharpoons N$' and goes on to say that 'information can be changed into negentropy and vice versa' (*loc. cit.*, p. 153). This is the statement of the Negentropy Principle of Information, and is regarded by Brillouin as having widespread validity. For example (*loc. cit.*, p. xii), he says it is 'justified by a variety of examples ranging from theoretical physics to everyday life'.

The reader is left with the impression that what is being asserted is the interconvertibility of thermodynamic entropy and information, with appropriate change of sign, in a quite general sense. In fact, however, the only example actually dealt with by Brillouin, either in his book or in his research papers, is that of Maxwell's Demon. So let us turn to his treatment of the Demon and we shall proceed to show that it is fallacious to suppose that it implies any quantitative relationship between entropy and information.

The credit for first effectively solving the problem of Maxwell's Demon is due to Demers (1944). In two successive papers he made very detailed energy and entropy balances on the Maxwell set-up, allowing for factors such as the Brownian motion of the trap door. What was to prove the most significant factor was the necessity for the Demon to make use of some *physical means* of distinguishing between the gas molecules in advance of selecting them to pass, or not to pass, through the trap door. Typically, the gas molecules have to be illuminated by quanta of radiation of energy $h\nu$ which sufficiently exceeds the energy of the black body radiation within the enclosure.

Brillouin picked up on this important point and simplified Demer's earlier discussion. He supposed that the Demon must be provided with something like an electric torch whose filament is at a temperature $T_f$ and gives photons of energy $h\nu_f > kT_0$, where $T_0$ is the temperature of the enclosure.

'The torch,' he wrote, 'is a source of radiation not in equilibrium. It pours negative entropy into the system. From this negative entropy the Demon obtains information. Using this information, he operates the trap door and rebuilds negative entropy, thus completing a cycle:

$$\text{negentropy} \rightarrow \text{information} \rightarrow \text{negentropy.'}$$

This, we suggest, is deceptive. Let us do *exactly the same arithmetic* as is done by Brillouin but expressing the matter entirely in thermodynamic terms and without reference to 'information'.

Consider first an energy transfer, $E$, from the filament, at temperature $T_f$, to the gas, at temperature $T_0$, during a certain time. The entropy production is

$$\Delta S_{\text{et}} = \frac{E}{T_0} - \frac{E}{T_f} > 0 \quad \text{since } T_f > T_0 \tag{5.2}$$

where the subscript 'et' is for energy transfer. Now consider the Demon regarded as a physical body. As Brillouin puts it: 'he can detect a molecule when at least one quantum of energy $hv_f$ is scattered by the molecule and absorbed in his eye ....' The increase in the entropy of the Demon's body, which is at the temperature $T_0$ of the enclosure, is

$$\Delta S_D = hv_f / T_0$$

If the Demon selects that molecule for passage through the trap door this can *diminish* the entropy of the gas. This effect will be of the order:

$$\Delta S_g = k \ln (W_0 - \delta W)/W_0$$

where $W_0$ is the original number of micro-states of the gas, in both compartments, and $W_0 - \delta W$ is the number of micro-states after the segregation of the selected molecule in, say, the right-hand compartment. Hence

$$\Delta S_D + \Delta S_g = hv_f/T_0 + k \ln (W_0 - \delta W)/W_0$$
$$\approx k \left( \frac{hv_f}{kT_0} - \frac{\delta W}{W_0} \right)$$
$$> 0 \tag{5.3}$$

since $\delta W/W_0 \ll 1$ and $hv_f/kT_0 > 1$.[55] Thus we conclude, as does Brillouin, that the Demon cannot infringe the Second Law.[56]

The point of our paraphrase of Brillouin's argument, using exactly the same factors, is simply to show that there is no necessity whatsoever for the process to be conceived as involving successive stages in which negentropy is converted into 'information' and then back into negentropy. All that is required for dealing with the physics of the situation is to make the summation of four entropy terms, two of them in (5.2) and the other two in (5.3), all of them thermodynamic entropy. This indeed is just how one deals with many other physico–chemical problems where one or more negative entropy changes are balanced by one or more positive entropy changes, resulting in an overall change which is positive or zero in accordance with the Second Law.

What can be conceded is that the general ideas of information theory have been useful in this example in a purely heuristic sense. They provided the initial insight that the Demon must be able to distinguish between the gas molecules by physical means. In that respect those ideas can truly be said to have contributed to the *process of solving* the problem of Maxwell's Demon; but once the essential insight has been obtained the solution can be carried through without any quantitative use of information theory. All the factors are capable of being dealt with by straight thermodynamics.

The reason why we have given some space to the Negentropy 'Principle' is that a number of scientists have regarded it as having great significance and have compared it with Aristotelian ideas about the relation of the world to the knowing subject. Let us therefore pursue the matter a little further.

There is, we think, nothing problematic about the idea that the making of a measurement on a physical system normally entails an entropy increase. As long ago as 1929, Szilard argued that the act of making a binary measurement (e.g. in deciding that some coordinate has a value closer to $+x$ than to $-x$) requires an entropy production of at least $k \ln 2$. For example, suppose it is required to ascertain the position of an ammeter needle; the needle has an average Brownian energy equal to $kT$ and therefore more than this amount of energy must be transferred to the needle in order to obtain a 50 per cent chance that an observation is correct, and is not simply due to the Brownian motion. Thus the entropy created in the ammeter exceeds $kT/T = k$. Gabor and Brillouin, in detailed studies, established $k \ln 2$ as the absolute minimum entropy production in a binary measurement, in agreement with Szilard's conjecture.

What *is* problematic is the other half of Brillouin's thesis in his 'Principle', i.e. it is the assertion that 'information', once it has been obtained, can then be used for achieving the reverse process – that of *reducing* the entropy of some physical system.

Surely there must be some restriction! For instance, it is entirely outside our experience that a sequence of symbols on a piece of paper or on a tape, or held in someone's head, can ever act as the physical *cause* (as distinct from an instruction about what to do) of an entropy reduction in some entirely separate system. What is clearly needed is a statement of the necessary and sufficient conditions under which Brillouin's interconvertibility thesis will hold, but this he did not provide.

In the context of the Demon, one obviously necessary condition relates to time. For unless the Demon opens the trap door quickly enough, when he 'sees' the hot molecule approaching, his efforts will be a failure. The same will apply to any device whose object is to seize hold of the momentary fluctuations which occur in physico–chemical systems. That device and the system in question must operate on each other within a time-like interval which is short in comparison with the relaxation time of the fluctuations. Yet this is not also a sufficient condition. That this is so, follows from the fact that it was possible to deal with the Demon problem in purely thermodynamic terms and without invoking the interconvertibility thesis.

Sometimes the system in question may be said to have 'memory' – or in less anthropomorphic language to have *correlations* between its parts. Such is the case in the spin echo system which was referred to in § 3.1 and § 3.4. We shall turn to a more detailed discussion of fluctuations and correlations in the next section.

By way of a brief summary, the Negentropy 'Principle' is very far from being a law of nature. In reply to our own question about necessary and sufficient conditions we suggest that interconvertibility holds if, and only if, an item, $\Delta H$, of supposed 'information' can be expressed *identically* as a heat quantity and a temperature according to the relation

$$\Delta H = \int (\mathrm{d}q/T)_{\mathrm{rev.}}$$

That is to say, when 'information' is *the same* as thermodynamic entropy and when the former notion is therefore superfluous.[57]

It is of interest that Jaynes' school of information theory makes singularly little mention of Brillouin's thesis. Indeed, Jaynes himself (1963) expressed what may be a warning. '... we have to emphasize,' he wrote, that Shannon's measure and the thermodynamic entropy 'are entirely different concepts. Our job cannot be to *postulate* any relation between them; it is rather to *deduce* whatever relations we can from known mathematical and physical facts.' Further discussion is to be found in papers and books by Tisza (1966), Watanabe (1969), Penrose (1970), Jauch and Báron (1972), Lindblat (1973, 1974), Costa de Beauregard and Tribus (1974) and Skagerstam (1975).

## 5.5. Fluctuations and correlations

Although thermodynamic entropy is an extensive quantity the same is not necessarily exactly true of the statistical analogue $S_G$, the Gibbs' entropy. In the following, it will be shown how small deviations from the additivity of $S_G$ can be taken care of within the structure of statistical mechanics and without reference to information theory. First we deal with natural fluctuation phenomena and then turn to processes of interaction and correlation.

In §4.2 we were concerned with the 'mixing' of $N$ atoms of a monatomic gas occupying a volume $V$ with an equal amount and volume of the same gas. It was pointed out in note 36 that if the term $\ln (2\pi N)^{1/2}$ had been included in the Stirling approximation, as strictly speaking it should have been, this would have resulted in a very small deviation from the additivity of the Gibbs' entropy. The deviation is positive and is given by:

$$\Delta S_G(N) = k \ln (\pi N)^{1/2} \tag{5.4}$$

where $(N)$ denotes that $N$ is fixed. The origin of this deviation is that fluctuations of larger magnitude become possible as soon as the partition is removed and thus give rise to additional quantum states.

Let us continue with this example. It will be supposed that the gaseous system is in thermal equilibrium with a heat reservoir and that after 'mixing' of the $2N$ atoms has occurred a partition is pushed down so that the volume $2V$ is divided into two equal parts $A$ and $B$. In general, it will not happen that exactly $N$ atoms are trapped on each side. Let $P(N_A)$ be the probability that $N_A$ atoms are trapped in $A$, so that $P(N_A)$ is also the probability that $N_B = 2N - N_A$ atoms are trapped in $B$. Denote by $p_i^A(N_A)$ and $p_j^B(N_A)$ the conditional probabilities that $A$ and $B$ are in their *i*th and *j*th quantum states respectively given that there are $N_A$ atoms in $A$. Also denote by $p_{ij}^C(N_A)$ the conditional probability of the joint event. This can be written as the product

$$p_{ij}^C(N_A) = p_i^A(N_A) \cdot p_j^B(N_A) \tag{5.5}$$

because when the combined system, $C$, is in the heat reservoir there are no mutual restrictions on the energies of $A$ and $B$, and there is no other reason why the states of $A$ and $B$ should be correlated.

Now the unconditional probability of the joint event is

$$p_{ij}^C = p_{ij}^C(N_A) \cdot P(N_A) \tag{5.6}$$

and the Gibbs' entropy of the whole system, $C$, is therefore given by

$$S_G^C/k = -\sum_{i,j} p_{ij}^C \ln p_{ij}^C$$

$$= -\sum_{i,j,N_A} p_{ij}^C(N_A)P(N_A) \ln p_{ij}^C(N_A) - \sum_{i,j,N_A} p_{ij}^C(N_A)P(N_A) \ln P(N_A)$$

$$= -\sum_{N_A} P(N_A) \sum_{i,j} p_{ij}^C(N_A) \ln p_{ij}^C(N_A) - \sum_{N_A} P(N_A) \ln P(N_A) \tag{5.7}$$

In obtaining the last line of (5.7) it will be noted that $\sum_{i,j} p_{ij}^C(N_A) = 1$ for any particular $N_A$. Substituting (5.5) into (5.7) then gives[58]

$$S_G^C = \sum_{N_A} P(N_A)\{S_G^A(N_A) + S_G^B(N_A)\} - k \sum_{N_A} P(N_A) \ln P(N_A) \tag{5.8}$$

where $S_G^A(N_A)$ and $S_G^B(N_A)$ are the Gibbs' entropies of $A$ and $B$ respectively for a given number $N_A$ of atoms in $A$. If we consider a large number of successive openings and closings of the partition the last equation can be written

$$S_G^C = \overline{S_G^A(N_A)} + \overline{S_G^B(N_A)} - k \overline{\ln P(N_A)} \tag{5.9}$$

where the first two terms on the right-hand side are the weighted mean values of the entropies of parts $A$ and $B$ respectively after the closings and $\overline{\ln P(N_A)}$ is the corresponding weighted mean value of $\ln P(N_A)$. Since the latter is a negative quantity, it follows that $S_G^C$ slightly exceeds the sum of the mean entropies of parts $A$ and $B$. The physical reason for this slight departure from strict additivity has been mentioned already.

The values of $\ln P(N_A)$ form a random series over a number of closings, due to the chance character of the fluctuations in the number of atoms in parts $A$ and $B$. Now the entropies $S_G^A(N_A)$ and $S_G^B(N_A)$ only have meaning when the shutter is closed whereby $N_A$ takes on a definite value. $S_G^C$, on the other hand, remains constant at a value, as follows from (5.4), equal to $S_G^C(0) + k \ln (\pi N)^{1/2}$. Here, $S_G^C(0)$ is the value at the initial instant *before* the partition was first opened when $N_A$ was exactly equal to $N_B$.

The value of $-k \overline{\ln P(N_A)}$ is a little greater than $k \ln (\pi N)^{1/2}$ – but only very slightly greater since $N_A = N_B = N/2$ remains the 'most probable' value of $N_A$.

The fluctuations occur, of course, without the need for making any measurements on the system; it is simply a matter of 'trapping' the fluctuations by closing the partition at random times. On the other hand, if the number of atoms, $N_A$, were somehow to be counted, the corresponding probability, $P(N_A)$, would become unity. Thus (5.8) would become $S_G^C(N_A) = S_G^A(N_A) + S_G^B(N_A)$ where $S_G^C(N_A)$ is less than $S_G^C$ by an amount equal to about $k \ln (\pi N)^{1/2}$ with great probability. This minute reduction in the entropy of the combined system due to counting will be compensated for by an at least equal increase elsewhere. In fact, according to what was said in § 5.4, based on Brillouin and Gabor, the counting of $N_A$ atoms may be expected to require an entropy increase of about $kN_A \ln 2$, and this is much greater than $k \ln (\pi N)^{1/2}$.

It remains true, however, that one might be lucky, once in a zillion years, and close the partition when a really significant fluctuation had occurred. The resulting pressure difference, between parts $A$ and $B$ of the system, could then be used to obtain mechanical work at the expense of heat taken in by the system from the heat reservoir. The entropy decrease in the reservoir might then be expected to be much greater, on this highly exceptional occasion, than the entropy increase entailed by the simple pressure measurement which would be sufficient to establish that the large fluctuation had indeed occurred.

This again emphasises that the classical Second Law is true only on the average; very rare occasions will arise when the Law is slightly falsified. The question arises whether it can be re-formulated in such a way that it retains its status as an 'impossibility statement'. One such formulation is due to Jaynes (1963): 'Spontaneous decreases in the entropy, although not absolutely prohibited, cannot occur in an experimentally reproducible process.' Yet another formulation was put forward by Penrose (1970, p. 222), who also discussed fluctuations in much greater detail than has been done above.

We turn now to a different situation where there are interactions between two or more parts of a system resulting in correlations. The following theorem is based on Gibbs (1902, Chapter XI, Theorem VIII). This has been used extensively in information theory but is presented below in statistical mechanical terms. For simplicity we consider only two sub-systems, $A$ and $B$, and initially they may be physically apart or they may consist of distinct species of particles which have not yet interacted.

Let (0) stand for the state of affairs before interaction and let ($t$) refer to a time when interaction has occurred. Let $p_{A_i}(0)$ and $p_{B_j}(0)$ be the initial probabilities that $A$ is in its $i$th quantum state and that $B$ is in its $j$th, respectively. Since these probabilities are independent, we can write for the joint probability

$$p_{A_i B_j}(0) = p_{A_i}(0) \cdot p_{B_j}(0) \tag{5.10}$$

Also

$$p_{A_i}(0) = \sum_{B_j} p_{A_i B_j}(0) \tag{5.11}$$

$$p_{B_j}(0) = \sum_{A_i} p_{A_i B_j}(0) \tag{5.12}$$

Denoting by $S_G^C$ the Gibbs' entropy of the composite system, it follows that:

$$S_G^C(0) = S_G^A(0) + S_G^B(0) \tag{5.13}$$

Thus before interaction occurs the Gibbs' entropy is additive.

After interaction has taken place, the analogue of (5.10), with ($t$) in place of (0), will not hold since the sub-systems are no longer independent. On the other hand, the analogues of (5.11) and (5.12) may be assumed still to apply – for example, if $A$ and $B$ are separated after interaction. We can then obtain:

$$S_G^A + S_G^B - S_G^C = -k \sum_{A_i} p_{A_i} \ln p_{A_i} - k \sum_{B_j} p_{B_j} \ln p_{B_j}$$

$$+ k \sum_{A_i \cdot B_j} p_{A_i \cdot B_j} \ln p_{A_i, B_j}$$

$$= -k \sum_{A_i \cdot B_j} p_{A_i \cdot B_j} \ln (p_{A_i} \cdot p_{B_j})$$

$$+ k \sum_{A_i \cdot B_j} p_{A_i \cdot B_j} \ln p_{A_i \cdot B_j} \qquad (5.14)$$

where the notation ($t$) has not been included for simplicity.

It then follows from the same familiar inequality as is used in Appendix 3.2 that

$$S_G^A(t) + S_G^B(t) \geqslant S_G^C(t) \qquad (5.15)$$

Thus additivity no longer holds. It will be noticed that the sign of the inequality is the opposite of that which occurred in (5.9), indicating an important difference between the phenomena of fluctuations and of interactions. It may be said that the effect of the correlations in the composite system $C$ is to reduce the entropy below the value which it would otherwise have had. Nevertheless, there will, in general, be an overall entropy increase:

$$\left.\begin{array}{c} S_G^C(t) \geqslant S_G^C(0) \\ S_G^A(t) + S_G^B(t) \geqslant S_G^A(0) + S_G^B(0) \end{array}\right\} \qquad (5.16)$$

since the interaction may occur irreversibly.

A more detailed discussion of the inequality (5.15) is available in Yvon (1969, Chapter 3), Penrose (1970, p. 216) and Landsberg (1978, p. 130). Yvon defines the *correlation entropy* as being the negative quantity $S_G^C - (S_G^A + S_G^B)$ as obtained from (5.15). He also points out that the validity of (5.15) depends on the assumption that the interaction does not alter the classical phase variables with the consequence that the phase space of the composite system, $C$, is the product of the phase spaces of $A$ and $B$. An equivalent assumption is implicit in the foregoing discussion of quantum states.

It is particularly to be noticed that the concept of correlation must be understood as extending, in the classical case, over the momentum space as well as over the coordinate space. If the correlations were to be taken, mistakenly, as signifying a greater degree of 'orderliness' in physical space, one might be led into the same kind of error about the interpretation of entropy as was discussed in §2.5 in connection with the crystallisation of a super-cooled melt.

The concept of correlations is, of course, well established in physical science, especially in regard to the pairwise correlation of molecules as in Boltzmann's original *H*-theorem. Correlation techniques have been adopted by many scientists, notably by Prigogine and his colleagues (1973). They have been particularly useful in achieving an understanding of certain phenomena such as the 'spin echo' which was discussed in Chapter 3 and in Appendix 3.1. In this instance, the existence of correlations is the source of an apparent Loschmidt-type reversal – and yet this is not contrary to the Second Law due to the occurrence of other entropy producing processes within the system.

### 5.6. Conclusions

Although information theory is more comprehensive than is statistical mechanics, this very comprehensiveness gives rise to objectionable consequences when it is applied in physics and chemistry. Since the subjective interpretation of probability, on which information theory has been based, is more general than is the objective interpretation, the language in which the theory is presented necessarily takes on a subjective character. A probability distribution is thus said to represent *our* state of knowledge, whereas in physico–chemical contexts it is usually justifiable to go much further in an objective direction and to regard a probability distribution as being characteristic of the nature of the system in question and of the physical constraints upon it.

It remains true, nevertheless, that information theory can be of value in an heuristic sense. This was seen to be so in the Maxwell Demon problem and it is also exemplified in cosmology by some of the papers of Bekenstein (1973), Layzer (1976) and Hawking (1976). Notions about 'loss of information' can sometimes be intuitively useful. But they can also, like the comparable concept of 'disorder', give rise to mistakes. Great care has to be exercised. It needs to be kept in mind that *thermodynamic* entropy is fully objective, as was shown in Chapter 1, and the same must apply to any other 'entropy' which is used as a surrogate. Indeed, our general argument in this chapter is that, in all physico–chemical contexts, it must be examined whether or not the so-called 'information entropy' is identical with an item of thermodynamic entropy, as specified for a system at equilibrium and substantially isolated, or in a heat bath.

Let us end with two or three more general remarks. In saying *substantially* isolated, we draw attention once again to the fact that physico–chemical systems can never be totally isolated; they are always subject to external influences such as those which are due to minutely changing gravitational fields. The effect is to destroy the correlations and to disrupt the deterministic temporal development of the system in question.

In certain respects the concept of *irreversibility* is more fundamental than is the concept of entropy, and is also more widely applicable. In general, one can make clear and unambiguous statements about entropy only in connection with laboratory systems having prescribed constraints. As regards, say, the Sun and stars, it may not be known whether their entropy is increasing or decreasing since they are open systems. But it can certainly be said of those bodies that they are the seats of irreversible processes on an enormous scale. It is the nucleo-synthesis in the Sun which is the source of the free energy which allows us to 'prepare' non-equilibrium systems in the laboratory for which entropy increase can be demonstrated. Furthermore, the irreversibility within the cosmos is fully objective in the sense of being independent of man's presence.

Finally, and in spite of the position which has been adopted in this book, it remains the case, of course, that what can be said to be objective in science can usually be claimed to be so only within the context of some particular *theory*. This raises the question: How does the matter stand in the light of the fact that theories are human creations?

To attempt to deal with this at all adequately would go far beyond our limited scope, but it may be suggested that objectivity$_2$ only makes sense when one adopts a belief in *realism*, such as was briefly discussed in §1.2. One has to suppose that scientific theories are at least conjectural approximations to what the world may really be like, and that the theories, although created by thought, mirror a reality which is independent of thought.

# *The quantum axioms*

The purpose of this appendix is to fill out, especially in the mathematics, the survey of quantum mechanics in §2.2. We first remind the reader about the meaning of vectors in the three-dimensional space which is familiar. Consider a set of three axes $OX_1, OX_2$ and $OX_3$ at right angles to each other and intersecting at the origin. Suppose one can go from point $P$ to point $Q$ first by a displacement of length $x_1$ parallel to $OX_1$, then by a displacement of length $x_2$ parallel to $OX_2$, and finally by a displacement of length $x_3$ parallel to $OX_3$:

$$PQ = x_1 OX_1 + x_2 OX_2 + x_3 OX_3$$

where $OX_1$, $OX_2$ and $OX_3$ are *unit vectors*. Since the order of the displacements is of no significance, there are further relationships of the type:

$$x_2 OX_2 + x_1 OX_1 = x_1 OX_1 + x_2 OX_2$$

The position of any point $P$ is defined by the vector $OP$. If

$$OP = x_1 OX_1 + x_2 OX_2 + x_3 OX_3$$

then the three-tuple $(x_1, x_2, x_3)$ is a *representation* of $OP$. Using a different set of perpendicular vectors of unit length $OX'_1, OX'_2, OX'_3$ through the same origin:

$$OP = x'_1 OX'_1 + x'_2 OX'_2 + x'_3 OX'_3$$

and $(x'_1, x'_2, x'_3)$ is a different representation of the vector $OP$ while the latter is unchanged. $OX_1, OX_2, OX_3$ provide an *orthonormal basis set* and one can evidently change to a different orthonormal basis set and obtain an alternative representation of $OP$.

The same ideas apply to spaces of dimensionality other than three and even to a space of infinite dimensions. The dimensionality of the space is the smallest number of axes required to express any vector $OP$ as $\sum_j x_j OX_j$. The vectors $OX_1$, $OX_2$, etc., are now more conveniently denoted as $|1\rangle, |2\rangle$, etc. The Hilbert space, $\Omega$ (sometimes called a separable Hilbert space), and various things concerning it, is defined below. It is equivalent to the Hilbert space of von Neumann (1955), although he used a different set of axioms.

1. The Hilbert space, $\Omega$, is the set of all vectors $|\psi\rangle$ which can be written as the expansion

$$|\psi\rangle = \sum_j \psi_j |j\rangle$$

where each $\psi_j$ may be any complex number and is called the *coefficient* of $|j\rangle$. The number of $|j\rangle$ may be infinite but it is always possible to put the $|j\rangle$ into a sequence like $|1\rangle, |2\rangle, |3\rangle$, etc. When the number of $|j\rangle$ is infinite, the $\psi_j$ must satisfy the condition that $\sum_j |\psi_j|^2$ converges to a finite limit. $\{\sum_j |\psi_j|^2\}^{1/2}$ is called the *length* of $|\psi\rangle$.

2. Suppose $|\psi\rangle = \sum_j \psi_j |j\rangle$ and $|\phi\rangle = \sum_j \phi_j |j\rangle$ and both belong to $\Omega$. The sum of $|\psi\rangle$ and $|\phi\rangle$ is written $|\psi\rangle + |\phi\rangle$ and is defined by

$$|\psi\rangle + |\phi\rangle \equiv \sum_j (\psi_j + \phi_j)|j\rangle$$

It is easily shown that the length of $|\psi\rangle + |\phi\rangle$ is less than or equal to the length of $|\psi\rangle$ plus the length of $|\phi\rangle$ so that $|\psi\rangle + |\phi\rangle$ belongs to $\Omega$.

It follows from this definition that

$$|\phi\rangle + |\psi\rangle = |\psi\rangle + |\phi\rangle$$

3. If $c$ is any complex number and $|\psi\rangle$ is any member of $\Omega$ then

$$c|\psi\rangle \equiv \sum_j c\psi_j |j\rangle$$

$c|\psi\rangle + d|\phi\rangle$ is sometimes written as $|c\psi + d\phi\rangle$.

4. If $|\psi\rangle$ and $|\phi\rangle$ belong to $\Omega$, the *inner product* is defined as

$$\langle \phi | \psi \rangle = \sum_j \phi_j^* \psi_j \tag{1}$$

The inner product always exists since it can be shown to be less than or equal to the length of $|\psi\rangle$ times the length of $|\phi\rangle$. If $\langle \phi | \psi \rangle = 0$, $|\phi\rangle$ and $|\psi\rangle$ are said to be *orthogonal*.

This completes the basic definitions and a number of deductions can now be made. The first is:

$$\langle \psi | \phi \rangle = \langle \phi | \psi \rangle^*$$

Also if $|\chi\rangle$ belongs to $\Omega$, then for any complex numbers $a, b$

$$\langle \phi | a\psi + b\chi \rangle = a\langle \phi | \psi \rangle + b\langle \phi | \chi \rangle$$
$$\langle a\psi + b\chi | \phi \rangle = a^*\langle \psi | \phi \rangle + b^*\langle \chi | \phi \rangle$$

$\langle \psi | \psi \rangle^{1/2}$ is the length of $|\psi\rangle$ and is zero if and only if each $\psi_j = 0$. In this case it is said that $|\psi\rangle = |0\rangle$, the *null vector*.

The Kronecker $\delta_{ij}$ is defined so that

$$\delta_{ij} \equiv \begin{cases} 1 & \text{if } i=j \\ 0 & \text{if } i \neq j \end{cases}$$

By choosing $\phi_j = \delta_{ij}$ in equation (1) we have $\langle i | \psi \rangle = \psi_i$. Therefore $\langle i | \psi \rangle$ is the coefficient of $|i\rangle$ in the expansion of $|\psi\rangle$ and, if $\langle i | \psi \rangle$ is known for each $i$, $|\psi\rangle$ is determined uniquely. By choosing $\phi_k = \delta_{ik}$ and $\psi_k = \delta_{jk}$ so that $|\phi\rangle = |i\rangle$ and $|\psi\rangle = |j\rangle$ and then applying equation (1)

$$\langle i | j \rangle = \delta_{ij}$$

This means that each $|i\rangle$ is of unit length and orthogonal to every other $|j\rangle$ so that it is said that $\{|j\rangle\}$, the set of all $|j\rangle$, provides an orthonormal basis set. Just as in the case of three dimensions, an alternative orthonormal basis set $\{|j'\rangle\}$ can be chosen such that each $\langle i' | j' \rangle = \delta_{ij}$ and such that *any* $|\psi\rangle$ in $\Omega$ can be expressed as $\sum_j \psi'_j |j'\rangle$. Thus both $\{|j\rangle\}$ and $\{|j'\rangle\}$ are said to be *complete*, or alternatively that they *span* the space $\Omega$.

If $\quad |\phi\rangle = \sum_j \phi'_j |j'\rangle \quad$ then

$$\langle \phi | \psi \rangle = \sum_{ij} \phi'^{*}_i \psi'_j \langle i' | j' \rangle = \sum_j \phi'^{*}_j \psi'_j$$

It follows that the properties of inner products do not depend on which orthonormal basis set is used. $\{\psi_j\}$ is the representation of $|\psi\rangle$ using the basis set $\{|j\rangle\}$. If one uses the other basis set the representation of $|\psi\rangle$ becomes $\{\psi'_j\}$ although $|\psi\rangle$ itself does not change.

Let us now turn to *operators* in a Hilbert space. Because $\langle \phi | \psi \rangle$ is always a scalar, one can think of $\langle \phi|$ as a functional which operates on any $|\psi\rangle$ to produce a scalar. Thus the $\langle \phi|$ form a space which, following Dirac (1958), is called a *bra* space while the $|\psi\rangle$ form a *ket* space. A bra vector and a ket vector together form a bracket, or inner product, and this is the reason for their names. Using the foregoing results:

$$\langle \phi | \psi \rangle = \sum_i \phi^{*}_i \psi_i = \sum_i \phi^{*}_i \langle i | \psi \rangle$$

Since this is true for any $|\psi\rangle$, one can conveniently write

$$\langle \phi | = \sum_i \phi^{*}_i \langle i|$$

If $\langle \phi | i \rangle$ happens to be known for each $|i\rangle$ then so is $\phi^{*}_i$ and the bra $\langle \phi|$ is uniquely determined.

Although a bra vector is a scalar-valued operator, what is usually meant by an operator in quantum mechanics is something which can act on a ket vector in $\Omega$ to produce another ket vector in $\Omega$. An operator is always placed on the left of a ket vector. For example, if $\hat{A}$ is an operator, a typical equation is $\hat{A}|\psi\rangle = |\phi\rangle$. If, in addition, $\hat{B}|\phi\rangle = |\chi\rangle$ it is said that $\hat{B}\hat{A}|\psi\rangle = |\chi\rangle$, thus defining the operator $\hat{B}\hat{A}$. It follows that $(\hat{C}\hat{B})\hat{A} = \hat{C}(\hat{B}\hat{A})$, the associative law. On the other hand, the commutative law only applies in special cases and in general $\hat{B}\hat{A} \neq \hat{A}\hat{B}$.

It is to be noted that $\hat{A}|\psi\rangle$ is not necessarily defined for every $|\psi\rangle$ in $\Omega$ and the set of $|\psi\rangle$ for which $\hat{A}|\psi\rangle$ is defined is called the *domain* of $\hat{A}$. In quantum mechanics the domains of most operators used are smaller than the complete Hilbert space. In the following it will be assumed, however, that at least one orthonormal basis set can be found which lies in the domains of those operators that are used. In short, no other basis sets are adopted.

$\hat{A}$ is said to be a *linear operator* if, for any $|\psi\rangle$ and $|\phi\rangle$ belonging to the domain of $\hat{A}$,

$$\hat{A}|\psi + \phi\rangle = \hat{A}|\psi\rangle + \hat{A}|\phi\rangle$$

All operators used in quantum mechanics are linear and an example is $|\phi\rangle\langle\chi| \equiv \hat{C}$, say. By definition, for any ket $|\psi\rangle$

$$\hat{C}|\psi\rangle \equiv |\phi\rangle\langle\chi|\psi\rangle$$

or, in other words, $|\phi\rangle$ multiplied by $\langle\chi|\psi\rangle$. The most general linear operator is $\sum_{ij} a_{ij}|i\rangle\langle j| \equiv \hat{A}$, say.

$$\langle k|\hat{A}|l\rangle = \sum_{ij} a_{ij}\langle k|i\rangle\langle j|l\rangle = a_{kl}$$

Hence, if a linear operator is such that $\langle i|\hat{A}|j\rangle$ is known for each $i, j$, $\hat{A}$ is uniquely determined. It is said that

$$\hat{A} = \sum_{ij} |i\rangle\langle i|\hat{A}|j\rangle\langle j| \tag{2}$$

$\hat{A}$ is said to be a *hermitian operator* if it is linear and if for any $|\psi\rangle$ in the domain of $\hat{A}$, $\langle\psi|\hat{A}|\psi\rangle$ is real. An equivalent definition of a hermitian operator is that for any $i, j$

$$\langle i|\hat{A}|j\rangle^* = \langle j|\hat{A}|i\rangle$$

or, alternatively, that for any $|\phi\rangle, |\psi\rangle$ in the domain of $\hat{A}$

$$\langle\phi|\hat{A}|\psi\rangle^* = \langle\psi|\hat{A}|\phi\rangle$$

Having defined the necessary mathematical formalism, the physical axioms of quantum mechanics can be put forward.

1. Any physical state of the system under consideration can be represented by a vector of unit length called the *state vector* in an appropriate Hilbert space. Conversely, if $|\psi\rangle$ is a vector of unit length in the appropriate Hilbert space then it represents a physical state of the system although it may be one of infinite energy.

2. To any physical observable $A$ there corresponds a unique hermitian operator which is denoted $\hat{A}$.

3. The result of any measurement of observable $A$ must be an eigenvalue of $\hat{A}$ although it is not usually possible to predict which one with certainty. It is also assumed here that $\hat{A}$ has a complete set of eigenvectors. Frequently, however, operators are used for which this is not the case. For example, $\hat{X}_1$, the operator which specifies the $OX_1$ component of the displacement of a particle from the origin, has no eigenvectors or eigenvalues in the proper sense. However, $X_1$ is an idealised observable which cannot be realised in practice since it implies a measurement of infinite accuracy. On the other hand, if a volume were divided into a finite number of regions a measurement disclosing in which region the particle lay would correspond to an operator with a complete set of eigenvectors. Each eigenvalue is the number of the region in which the particle lies and has an infinite degeneracy since there are an infinite number of states in each region.

4. This axiom has already been stated in the main text under 'Eigenvalues and eigenvectors'. It shows how the probability that an ideal measurement of $A$ gives $a_j$ is related to the eigenvectors of $\hat{A}$. It also deals with the state of the system after the measurement has been completed. It can be used to calculate the expectation value $\langle A \rangle$ of $\hat{A}$. $\langle A \rangle$ is the average result of a measurement of $A$ and is $\sum_j p_j a_j$ where $p_j$ is the probability that the result of the measurement is $a_j$, and is equal to $\sum_{i=1}^{d} |\langle \alpha_{ji} | \psi \rangle|^2$, as shown in the main text. This can be proved to equal $\langle \psi | \hat{A} | \psi \rangle$.

It may be noted that, if $|\psi\rangle$ is multiplied by $\exp i\theta$ for real $\theta$, each $p_{ij}$ is unaltered. This is the reason why $\exp i\theta |\psi\rangle$ represents the same physical state as $|\psi\rangle$.

If $\hat{A}$ and $\hat{B}$ are operators such that $\hat{A}\hat{B} = \hat{B}\hat{A}$, $\hat{A}$ and $\hat{B}$ are said to commute. If $\hat{A}$ and $\hat{B}$ do commute, it can be shown that there is a complete set of eigenvectors which are eigenvectors to both $\hat{A}$ and $\hat{B}$. This can be used to show that an ideal measurement of $A$ immediately followed by an ideal measurement of $B$ leaves the state of the system in an eigenvector of both $\hat{A}$ and $\hat{B}$ provided both measurements are instantaneous. A further measurement of $A$ or $B$ immediately following this leaves the state of the system unaltered. Let the results of the two measurements be $a_i$ and $b_j$. Since not all eigenvectors of $\hat{A}$ with eigenvalue $a_i$ are necessarily eigenvectors of $\hat{B}$ with eigenvalue $b_j$, the effect of doing the two measurements may tell us more about the final state of the system than just doing one measurement. If $\hat{A}$ and $\hat{B}$ commute, the observables $A$ and $B$ are said to be *compatible* because they can be measured simultaneously and the measurement of one does not affect the result of measuring the other. A *complete set* of observables $A$, $B$, $C$, etc., is such that all of the operators $\hat{A}$, $\hat{B}$, etc., commute with each other and such that a measurement of all of them is sufficient to establish the only possible state of the system. In other words, whatever $a_i$, $b_j$, $c_k$, etc., are found to be, there is only one eigenvector with this set of eigenvalues. Such a measurement is said to be *maximal.*

In the case where $\hat{A}$ and $\hat{B}$ do not commute there is no common set of eigenvectors. An ideal measurement of $A$ followed by an ideal measurement of $B$ puts the system first into an eigenstate of $A$ and then into an eigenstate of $B$. Thus $A$ is no longer determinate since the effect of the second measurement is that the result of the first no longer applies.

If the system does not happen to lie in a pure eigenstate of $\hat{A}$, the result of a measurement of $A$ is uncertain and a useful measure of the uncertainty of the result is the standard deviation of the result, $\sigma(A)$, say. $\sigma^2(A)$ is $\sum_j p_j(a_j - \langle A \rangle)^2$ which can be shown to be $\langle \psi | \hat{A}^2 | \psi \rangle - \langle \psi | \hat{A} | \psi \rangle^2$. Heisenberg proved the famous uncertainty relation

$$\sigma(A)\sigma(B) \geqslant \tfrac{1}{2}|\langle \psi | \hat{A}\hat{B} - \hat{B}\hat{A} | \psi \rangle| \tag{3}$$

An interesting special case of equation (3) occurs when $\hat{A}$ and $\hat{B}$ have some eigenvector $|\alpha\rangle$ in common. If $|\psi\rangle = |\alpha\rangle$, $\sigma(A)$ and $\sigma(B)$ are both zero. It is easy to see that the right-hand side of equation (3) is also zero.

5. $|\psi\rangle$ does not remain constant in time but varies according to the equation

$$i\hbar \frac{\partial}{\partial t} |\psi(t)\rangle = \hat{H}|\psi(t)\rangle \tag{4}$$

where $\hat{H}$ is the Hamiltonian operator. This equation – the Schrödinger time-dependent equation – is true even when $\hat{H}$ is allowed to vary with time. When $\hat{H}$ is constant we can take $|j\rangle$ as the time-independent eigenvector of $\hat{H}$ with eigenvalue $\varepsilon_j$. Writing $|\psi\rangle = \sum_j \psi_j(t)|j\rangle$ we obtain

$$i\hbar \sum_j \frac{\partial}{\partial t} \psi_j(t)|j\rangle = \sum_j \psi_j(t)\varepsilon_j|j\rangle$$

Hence
$$i\hbar \frac{\partial}{\partial t} \psi_j(t) = \varepsilon_j \psi_j(t)$$

Integrating this equation
$$\psi_j(t) = \exp(-i\varepsilon_j t/\hbar)\psi_j(0)$$

and hence
$$|\psi(t)\rangle = \sum_j \exp(-i\varepsilon_j t/\hbar)\psi_j(0)|j\rangle$$

as given in the main text.

6. This axiom states what operators can be used for the positions and momenta of particles. Consider a system of $n$ particles and let the position coordinates of the $j$th particle be $X_{j1}$, $X_{j2}$, $X_{j3}$ and let the corresponding momenta be $P_{j1}$, $P_{j2}$, $P_{j3}$. In quantum mechanics these quantities are observables which have appropriate operators. Let us express these operators in terms of the Schrödinger representation. Here the state $|\psi\rangle$ is represented by the wave function $\psi(x_{11}, \ldots, x_{n3})$ where the $x_{ij}$ are mathematical coordinates and have no direct physical meaning. The operator $\hat{X}_{j\mu}$ is now represented by $x_{j\mu}$ and the operator $\hat{P}_{j\mu}$ is represented by $-i\hbar \, \partial/\partial x_{j\mu}$. Thus $\hat{X}_{j\mu}|\psi\rangle$ is represented by

$$x_{j\mu}\psi(x_{11}, \ldots, x_{n3})$$

and $\hat{P}_{j\mu}|\psi\rangle$ is represented by

$$-i\hbar \, \frac{\partial\psi}{\partial x_{j\mu}}(x_{11}, \ldots, x_{n3})$$

From these relations one can show that $(\hat{X}_{11}\hat{P}_{11} - \hat{P}_{11}\hat{X}_{11})|\psi\rangle$ is represented by

$$x_{11}\left(-i\hbar \, \frac{\partial\psi}{\partial x_{11}}\right) + i\hbar \, \frac{\partial}{\partial x_{11}}(x_{11}\psi) = i\hbar\psi(x_{11}, x_{12}, \ldots)$$

Since $(\hat{X}_{11}\hat{P}_{11} - \hat{P}_{11}\hat{X}_{11})|\psi\rangle$ is thus represented by $i\hbar\psi(x_{11}, \ldots)$, it follows that

$$(\hat{X}_{11}\hat{P}_{11} - \hat{P}_{11}\hat{X}_{11})|\psi\rangle = i\hbar|\psi\rangle = i\hbar\hat{I}|\psi\rangle$$

where $I$ is the identity operator which transforms any vector into itself. This relation can be written

$$\hat{X}_{11}\hat{P}_{11} - \hat{P}_{11}\hat{X}_{11} = i\hbar\hat{I}$$

More generally, the following set of relations can be deduced:

$$\left.\begin{array}{l}\hat{X}_{j\mu}\hat{X}_{kv} - \hat{X}_{kv}\hat{X}_{j\mu} = \hat{P}_{j\mu}\hat{P}_{kv} - \hat{P}_{kv}\hat{P}_{j\mu} = 0 \\ \hat{X}_{j\mu}\hat{P}_{kv} - \hat{P}_{kv}\hat{X}_{j\mu} = i\hbar I\delta_{jk}\delta_{\mu v}\end{array}\right\} \qquad (5)$$

The only operators of this collection which do not commute are those operators $\hat{X}$ and $\hat{P}$ which have the same pair of subscripts. Having proved the set of relations (5) in the Schrödinger representation, it can be deduced that the same *commutation* relations apply in any representation.

Equation (3) can be applied to relations (5) to obtain

$$\sigma(X_{j\mu})\sigma(P_{kv}) \geqslant \tfrac{1}{2}\hbar\delta_{jk}\delta_{\mu v}$$
$$\sigma(X_{j\mu})\sigma(X_{kv}) \geqslant 0$$
$$\sigma(P_{j\mu})\sigma(P_{kv}) \geqslant 0$$

This means that it is possible to know simultaneously the exact positions of all $n$ particles. It is also possible to know their exact momenta simultaneously. It is possible to know the position of one particle and the momentum of another particle simultaneously. However, it is not possible to know the position and momentum of the *same* particle simultaneously.

7. This states how operators in quantum mechanics can be constructed to match certain observables. The classical equivalent to the observable is first expressed in terms of the $X_{j\mu}$ and $P_{k\mu}$. The quantum mechanics operators are obtained by substituting $\hat{X}_{j\mu}$ for $X_{j\mu}$ and $\hat{P}_{j\mu}$ for $P_{j\mu}$. Thus for a single particle if the Hamiltonian, $H$, is $(2m)^{-1}\{P_1^2 + P_2^2 + P_3^2\} + V(X_1, X_2, X_3)$ in classical mechanics, then in quantum mechanics

$$\hat{H} = (2m)^{-1}\{\hat{P}_1^2 + \hat{P}_2^2 + \hat{P}_3^2\} + V(\hat{X}_1, \hat{X}_2, \hat{X}_3) \tag{6}$$

Similarly, according to classical mechanics, the angular momentum, $L_3$, about the axis $OX_3$ is $X_1 P_2 - X_2 P_1$. In quantum mechanics the angular momentum operator is given by

$$\hat{L}_3 = \hat{X}_1 \hat{P}_2 - \hat{X}_2 \hat{P}_1$$

There are limitations to the application of this rule and, indeed, certain observables exist in quantum mechanics which have no classical equivalent, the most notable being electron spin.

When the Schrödinger representation is used and when there is just one particle moving in three dimensions, equations (4) and (6) can be combined to give

$$i\hbar \frac{\partial \psi}{\partial t}(x_1, x_2, x_3, t) = -\frac{\hbar^2}{2m}\left(\frac{\partial^2}{\partial x_1^2} + \frac{\partial^2}{\partial x_2^2} + \frac{\partial^2}{\partial x_3^2}\right)\psi(x_1, x_2, x_3, t)$$
$$+ V(x_1, x_2, x_3)\psi(x_1, x_2, x_3, t)$$

This is the well-known time-dependent Schrödinger equation for a single particle. To find the eigenstates and eigenvalues of the Hamiltonian one must solve the time-*in*dependent Schrödinger equation as follows, where $\varepsilon_j$ is the eigenvalue and $\psi_j(x_1, \ldots)$ represents the eigenvector:

$$\varepsilon_j \psi_j(x_1, x_2, x_3) = -\frac{\hbar^2}{2m}\left(\frac{\partial^2}{\partial x_1^2} + \frac{\partial^2}{\partial x_2^2} + \frac{\partial^2}{\partial x_3^2}\right)\psi_j(x_1, x_2, x_3)$$
$$+ V(x_1, x_2, x_3)\psi_j(x_1, x_2, x_3)$$

Although this completes our summary of the axioms, it may be added that a problem that is of great importance in statistical mechanics is about what the Hilbert space should be when two separate systems, $S_1$ and $S_2$, are thought of as constituting a single system, $S$.

Let $\Omega, \Omega_1$ and $\Omega_2$ be the Hilbert spaces of $S, S_1$ and $S_2$ respectively. If $S_1$ is in state $|\phi\rangle$ and $S_2$ is in state $|\psi\rangle$, it will be said that $S$ is in state $|\phi\rangle|\psi\rangle$, or more simply state $|\phi\psi\rangle$. The corresponding bra vector is $\langle\phi\psi|$. Suppose $\Omega_1$ is spanned by the basis set $|1_1\rangle, |2_1\rangle$, etc., and $\Omega_2$ is spanned by the basis set $|1_2\rangle, |2_2\rangle$, etc. The direct product of $\Omega_1$ and $\Omega_2$ is the Hilbert space denoted by $\Omega_1 \otimes \Omega_2$. It is defined so that the basis vectors of $\Omega_1 \otimes \Omega_2$ are the $|j_1 j_2\rangle$ for all possible $j_1, j_2$. Consequently if $\Omega_1$ and $\Omega_2$ are of finite dimensionalities, $m_1$ and $m_2$, then the dimensionality of $\Omega_1 \otimes \Omega_2$ is $m_1 m_2$. If $S_1$ is in state $|j_1\rangle$ and $S_2$ is in state $|j_2\rangle$, the probability that an ideal measurement will find $S_1$ to be in state $|i_1\rangle$ and $S_2$ to be in state $|i_2\rangle$ is zero unless $i_1 = j_1$ and $i_2 = j_2$. Applying equation (2.8) we obtain:

$$\langle i_1 i_2 | j_1 j_2 \rangle = \delta_{i_1 j_1} \delta_{i_2 j_2}$$

This is consistent with the assumption that the basis vectors of $\Omega_1 \otimes \Omega_2$ are orthonormal to each other.

If systems $S_1$ and $S_2$ do not interact with each other, then one would expect the Hilbert space of $S$ to be $\Omega_1 \otimes \Omega_2$. Let $H_1$, $H_2$ and $H$ be respectively the Hamiltonians of $S_1$, $S_2$ and $S$ and let $\hat{I}_1$ and $\hat{I}_2$ be the identity operators of $\Omega_1$ and $\Omega_2$ respectively. Then $\hat{H}_1 \otimes \hat{I}_2$ is the operator acting on $\Omega$ such that

$$\hat{H}_1 \otimes \hat{I}_2 |\phi\psi\rangle = |\hat{H}_1 \phi\rangle |\psi\rangle$$

for any $|\phi\psi\rangle$. Similarly

$$\hat{I}_1 \otimes \hat{H}_2 |\phi\psi\rangle = |\phi\rangle |\hat{H}_2 \psi\rangle$$

If the systems $S_1$ and $S_2$ do not interact then

$$\hat{H} = \hat{H}_1 \otimes \hat{I}_2 + \hat{I}_1 \otimes \hat{H}_2$$

If $|\phi\rangle$ is an eigenvector of $\hat{H}_1$ with eigenvalue $\lambda$ and $|\psi\rangle$ is an eigenvector of $\hat{H}_2$ with eigenvalue $\mu$ then $|\phi\psi\rangle$ is an eigenvector of $\hat{H}$ with eigenvalue $\lambda + \mu$. Very often it is more convenient to write just $\hat{H}_1$ for $\hat{H}_1 \otimes \hat{I}_2$ and similarly for $\hat{H}_2$ so that $\hat{H} = \hat{H}_1 + \hat{H}_2$.

If there is an energy interaction between $S_1$ and $S_2$ but the systems do not exchange particles, then $\Omega$ is still $\Omega_1 \otimes \Omega_2$. There is now an interaction Hamiltonian $H_{12}$ and

$$H = H_1 + H_2 + H_{12}$$

This means that the eigenvectors of $\hat{H}$ are not necessarily of the form $|\phi\psi\rangle$. Perturbation theory, however, indicates that when $H_{12}$ is small the eigenvectors and eigenvalues are not very different from the case when it is zero. When $H_{12}$ is small we have two weakly interacting systems and it is often a good approximation to assume the eigenvectors and eigenvalues are the same as when there is no interaction.

If the two systems $S_1$ and $S_2$ can exchange particles then $\Omega$ is not the same as $\Omega_1 \otimes \Omega_2$. The number of possible states of $S$ has been increased by allowing the exchange to occur, but the actual number depends on whether the particles have symmetric or anti-symmetric wave functions.

# *Equal* a priori *probabilities and random phases*

Grounds for accepting the basic postulate have been discussed by Ter Haar (1955) as well as by Tolman and others. A further argument in its favour will be put forward below. But it is important to notice that the postulate is not generally true and in particular does not apply to those systems which have not yet reached equilibrium, or to systems which are in *perfect* isolation, however long one waits. For suppose the state of the system at $t = 0$ is given by $|\psi\rangle = \sum_j a_j |j\rangle$ where the $|j\rangle$ are the eigenvectors of the Hamiltonian. In the case of a perfectly isolated system, the Hamiltonian is time independent and it has already been shown in §2.2.4 that $p_j$, the probability that an experiment will find the system in any state $|j\rangle$, does not change with time. Unless the system at $t = 0$ is in an exceptional condition such that the states of equal energy are already equally probable, such states will remain permanently *un*equally probable.

An attempt to justify the postulate in question requires a consideration of systems which are not quite perfectly isolated; and, of course, such systems are the rule rather than the exception. No real system is ever completely isolated since it is inevitably subject to small *random* perturbations such as may be due to minute variations in the gravitational field, penetration of the walls of the container by cosmic rays, and so on. These disturbances necessarily destroy the conditions under which the Schrödinger time-dependent equation would hold, and give rise to transitions between one energy state and another.

If such a system is already in approximate thermal equilibrium with the laboratory environment before it is isolated, it is known from experience that it will remain at almost perfectly constant temperature after the best isolation that can be achieved, since there is then little inward or outward leakage of heat. Any remaining perturbations, such as have been mentioned, can then be taken as random, and the range of the system's energy, $E$ to $E + \delta E$, will be very small.

Let $\hat{H}$ be the system's Hamiltonian when unperturbed and let $B$ be the set of eigenvectors of $\hat{H}$ whose eigenvalues lie in the foregoing range. Let $\hat{H}'(t)$

be the perturbation Hamiltonian due to random external influences. It is assumed that if $|i\rangle$ and $|j\rangle$ are eigenvectors of $\hat{H}$, $\langle i|\hat{H}'(t)|j\rangle$ is zero if $|i\rangle$ lies in $B$ and $|j\rangle$ does not, and vice versa for any value of $t$. This means that if initially the state of the system is a linear combination of eigenvectors within $B$ then this will always be so. We make the assumption that only transitions within the narrow range $E$ to $E+\delta E$ are possible, and that within this band there are transitions from any one state to any other. Such an assumption is consistent with perturbation theory. For in that theory the transition probability has a factor $(E_0-E_n)$ in the denominator, where $(E_0-E_n)$ is the energy difference between the initial state and the state numbered $n$. This implies that transitions between states of nearly equal energies are much more probable than transitions between states having a large energy difference.

Consider the particular case of a system, $S$, whose Hamiltonian changes at irregular intervals between two close lying Hamiltonians $\hat{H}_1$ and $\hat{H}_2$. The results which follow can be generalised to a system like that of the previous paragraph. Let the eigenstates and corresponding eigenvalues of $\hat{H}_1$ and $\hat{H}_2$ be respectively $|j_1\rangle, \varepsilon_{1j}, |k_2\rangle$ and $\varepsilon_{2k}$. If the system is in state $|\psi(t)\rangle$ at time $t$, $|\psi(t)\rangle$ can be expressed as either $\sum_j a_j(t)|j_1\rangle$ or as $\sum_k b_k(t)|k_2\rangle$. There is some unitary matrix, $L$, such that $b_k(t)=\sum_j L_{kj}a_j(t)$. If $\boldsymbol{a}(t)$ and $\boldsymbol{b}(t)$ are respectively the column matrices whose $j$th elements are $a_j(t)$ and $b_j(t)$, this relation can be expressed as

$$\boldsymbol{b}(t)=L\boldsymbol{a}(t)$$

Suppose that

$$\hat{H}(t)=\begin{cases}\hat{H}_1 & t<0 \\ \hat{H}_2 & 0\leqslant t\leqslant T \\ \hat{H}_1 & t>T\end{cases}$$

by which it is meant that $\hat{H}$ changes from $\hat{H}_1$ to $\hat{H}_2$ at $t=0$ and changes back to $\hat{H}_1$ at $t=T$.

$$b_k(T)=\exp\{-i\varepsilon_{2k}T/\hbar\}b_k(0)=d_{kk}b_k(0), \quad \text{say.}$$

The $d_{ij}$ form the elements of a diagonal matrix $D$ and the above relation can be expressed as

$$\boldsymbol{b}(T)=D\boldsymbol{b}(0)$$
$$\boldsymbol{a}(T)=L^{-1}DL\boldsymbol{a}(0)=M\boldsymbol{a}(0), \quad \text{say, where}$$
$$M=L^{-1}DL \tag{1}$$

If $T$ is a reasonably long interval and of *random* duration it may be assumed that the $d_{kk}$ are randomly placed on a unit circle except when there is degeneracy. If $\varepsilon_{2i}=\varepsilon_{2k}$ then, of course, $d_{ii}=d_{kk}$. It is also fair to assume the phase angles of the $a_j(0)$ are random if the Hamiltonian has been $\hat{H}_1$ for a reasonably long and random period of time. If $\hat{H}_1$ and $\hat{H}_2$ happen to have some eigenvectors in common, this argument breaks down, for if $|\psi(0)\rangle$ is such an eigenvector then the system will forever remain in that state. However, this is improbable and it is one of the assumptions of this argument that all states of $S$ are accessible under a succession of changes of the Hamiltonian. Since the phase angles of the $a_j(t)$ are assumed to be random, the probabilities

$$p_j(t) \equiv |a_j(t)|^2$$

are just as useful as the $a_j$ themselves. Letting $m_{jk}$ be the $jk$th element of $M$

$$p_j(T) = \left| \sum_k m_{jk} a_k(0) \right|^2 = \sum_k |m_{jk}|^2 p_k(0) + \sum_{k \neq l} m_{jk} m_{jl}^* a_k(0) a_l^*(0)$$

Since the phases of the $a_j(0)$ are assumed to be random, the second term of the right-hand side is a noise term whose average value is zero and which tends to randomise the $p_j(T)$. Let $\bar{p}_j(T)$ be the average value of $p_j(T)$ obtained by ignoring the noise term. Let $C$ be the matrix whose $jk$th element $C_{jk} \equiv |m_{jk}|^2$. Then writing $\bar{p}(T)$ for the column matrix whose $j$th element is $\bar{p}_j(T)$

$$\bar{p}(T) = C\bar{p}(0)$$

$$\sum_j C_{jk} = \sum_j |m_{jk}|^2 = (M \dagger M)_{kk}$$

Using the fact that $L\dagger L = 1$ and substituting equation (1), it follows that $M \dagger M$ is the identity matrix. In other words, $M$ is a unitary matrix. Hence $\sum_j C_{jk} = 1$. This is the condition that $C$ is a stochastic matrix and it follows that if $\sum_j \bar{p}_j(0) = 1$ then $\sum_j \bar{p}_j(T) = 1$, which is to be expected. In the same way the relation $\sum_k C_{jk} = 1$ can also be proved. Let $q$ and $\tilde{q}$ respectively be the column and row matrices whose elements are all 1. It follows from the fact that $\sum_k C_{jk} = 1$ that

$$Cq = q$$

and the fact that $\sum_j C_{jk} = 1$ that

$$\tilde{q}C = \tilde{q} \qquad (2)$$

If $\hat{H}$ changes to and fro between $\hat{H}_1$ and $\hat{H}_2$ a great many times after time $t$, the average value $\bar{p}_j(t)$ is given by

$$\bar{p}(t) = C_N C_{N-1} \dots C_2 C_1 \bar{p}(0)$$

The $C_j$ are all matrices with the properties of $C$ above but are not usually all equal to each other due to the fact that the time spent with Hamiltonian $\hat{H}_2$ usually varies with $j$. Below it is argued that, after a long time, when there are a large number of $C_j$, $\bar{p}(t)$ becomes $(1/N)\tilde{q}$ where $N$ is the total number of possible states. Let $w$ be a column matrix such that

$$\bar{p}(0) = \frac{1}{N}q + w \tag{3}$$

$w$ then represents the divergence between $\bar{p}(0)$ and a situation where all states are equally probable. Since $\sum_j \bar{p}_j(0) = 1, \tilde{q} \cdot \bar{p}(0) = 1$. Also $\tilde{q} \cdot q = N$. Hence from equation (3) $\tilde{q} \cdot w = 0$. In other words, $w$ lies in the space orthogonal to $q$. Now $C_j q = q$ for each $j$ so that

$$\bar{p}(t) = \frac{1}{N}q + C_N C_{N-1} \dots C_2 C_1 w$$

$$\tilde{q} \cdot Cw = \tilde{q} \cdot w \quad \text{by equation (2)}$$

$$= 0$$

Therefore $C_1 w$ lies in the same space as $w$ and is orthogonal to $q$. Labelling $C_1 w = w_1$ and $C_j w_{j-1} = w_j$ for each $j$ it follows by induction that $\tilde{q} \cdot w_j = 0$ for each $j$.

If $x$ is any column matrix, let $l(x) \equiv \sum_j |x_j|$. For any $x$, $l(Cx) \leqslant l(x)$ because

$$\sum_j \left| \sum_k C_{jk} x_k \right| \leqslant \sum_j \sum_k C_{jk} |x_k| = \sum_k \left\{ |x_k| \sum_j C_{jk} \right\} = \sum_k |x_k|$$

since each $C_{jk} \geqslant 0$ and $\sum_j C_{jk} = 1$.

If $l(Cx) = l(x)$ this means that for any $j$ the $C_{jk} x_k$ are either zero or of the same sign for all $k$. If $x$ is one of the $w_j$ then $\tilde{q} \cdot x = 1$ and some of the $x_k$ must be positive and some must be negative unless $w_j = 0$, in which case the theorem is already proved. It can be shown that, if $l(w_{j+1}) = l(w_j)$ for all $j$, not all states of $S$ would be accessible, contrary to hypothesis. In practice, due to the fact that $C$ contains the random matrix $D$, $l(w_{j+1})$ is always less than $l(w_j)$. One can say that for an average $j$,

$$l(w_{j+1})/l(w_j) < \eta \quad \text{where } \eta < 1$$

This means that as $n \to \infty, l(w_n) \to 0$. Therefore $w_n \to 0$ as $t \to \infty$ and $\bar{p}(t) \to (1/N)q$. Since the $p_j(t)$ differ from the $\bar{p}_j(t)$ only by the ignored noise terms, and since each $\bar{p}_j(t)$ converges to $1/N$, each $p_j(t)$ does so *at least* as fast.

This completes the proof that as time tends to infinity all states become equally probable. In the case where transitions are only allowed within a narrow energy band, the proof is exactly the same. Any state within this band, $B$, is accessible from any other state within $B$ by a series of transitions. All these states eventually become equally probable, justifying the postulate that states of equal energy are equally probable in a state of equilibrium.

# The derivation of the canonical distribution from the microcanonical distribution

In this Appendix we shall consider a system $S$ placed in thermal contact with a very much larger system $R$ and show that the probability, $p_\psi$, that $S$ lies in some energy eigenstate, $|\psi\rangle$, with eigenvalue $\varepsilon$, is proportional to $\exp -\beta\varepsilon$. Here $\beta$ is some property of $R$ which in the main text is identified with inverse temperature. In order to do this, two assumptions will have to be made. Assumption 1 says that the combined system, $S+R$, denoted by $T$, can be treated as a microcanonical ensemble. The energy eigenvectors of $T$ with eigenvalues between $U$ and $U+\Delta$, and all linear combinations of these vectors, form an $M$-dimensional space, $\Omega_B$, say. Assumption 1 asserts that $|\chi\rangle$, the state of $T$, must lie in $\Omega_B$ and is as likely to be any unit vector belonging to $\Omega_B$ as any other. This is equivalent to saying there are $M$ orthonormal energy eigenstates in $\Omega_B$ and the probability that a measurement finds $T$ to be in any one of these states is the same as any other and is $M^{-1}$. This is consistent with the assumption of equal *a priori* probabilities as discussed in Appendix 2.2. In reality, the assumption that $|\chi\rangle$ lies within a sufficiently narrow band for our arguments is implausible and later the argument will be generalised.

Let system $S$, taken in isolation, have Hamiltonian $H_S$ and let $|\psi\rangle$, a state of $S$, be an eigenstate of $H_S$ with eigenvalue $\varepsilon$. Similarly let $|\phi\rangle$, a state of $R$, be an eigenstate of $H_R$, the Hamiltonian of $R$ when taken in isolation, with eigenvalue $\eta$. $|\psi\phi\rangle$ is then that state of $T$ in which $S$ lies in $|\psi\rangle$ and $R$ lies in $|\phi\rangle$ (Appendix 2.1). Perturbation theory indicates that if the interaction Hamiltonian is small, any energy eigenstate of $T$ with energy $U$ is approximately a linear combination of states of the form $|\psi\phi\rangle$ such that $\varepsilon+\eta$ is close to $U$. This is rather more general than making the usual assumption that the energy states of $T$ are approximately of the form $|\psi\phi\rangle$ which is incorrect when degeneracy occurs. This result from perturbation theory leads to Assumption 2 which says $\Omega_B$ is that space which is spanned by all $|\psi\phi\rangle$ such that $U < \varepsilon+\eta < U+\Delta$. Since $\Omega_B$ is an $M$-dimensional space, Assumptions 1 and 2 mean that the probability that a measurement finds $T$ to be in any state $|\psi\phi\rangle$ satisfying the above conditions is $M^{-1}$.

Let $\{|j\rangle\}$ be any orthonormal basis set for the Hilbert space of $S$ such that each $|j\rangle$ is an eigenstate of $H_S$ with eigenvalue $\varepsilon_j$, say. Similarly, let $\{|i_R\rangle\}$ be any orthonormal basis set for $R$ where each $|i_R\rangle$ is an eigenstate of $H_R$ with eigenvalue $\eta_i$, say. It can readily be shown that there are $M$ orthonormal states of $T$ belonging to $\Omega_B$ of the form $|ji_R\rangle$. This set can be used as a basis set for $\Omega_B$ and will be denoted $B'$.

Having made these two assumptions we can now proceed with the proof. Let the number of orthonormal energy eigenstates of $R$ with eigenvalues between $\eta$ and $\eta+d\eta$ be $W_R(\eta)\,d\eta$. $W_R(\eta)$ is thus the energy distribution function and is not to be identified with the number $W$ of equation (2.14). Now $R$ can be regarded as being divided into a very large number $n$ of subsystems $\gamma_1, \gamma_2, \ldots, \gamma_n$ whose energy distribution functions are $w_1(\eta_1)$, $\ldots, w_n(\eta_n)$. Since $R$ is very large, each of these subsystems can be chosen as large and once again the interaction energy between them can be neglected. Thus

$$W_R(\eta) = \int w_1(\eta_1) \ldots w_{n-1}(\eta_{n-1})w_n(\eta-\eta_1-\eta_2-\cdots-\eta_{n-1})\,d\eta_1\,d\eta_2\ldots d\eta_{n-1}$$

In other words, $W_R$ is the convolution of $w_1 \ldots w_n$ where $n$ is large. From the mathematical properties of the convolution, it can be shown that $\log W_R(\eta)$, although not $W_R$ itself, is a slowly varying function of $\eta$. This enables $\log W_R$ to be expanded as a Taylor series and furthermore, provided $\varepsilon \ll U$ in equation (1) below, all but the first two terms in the expansion can be neglected. Thus

$$\log W_R(U-\varepsilon) = \log W_R(U) - \beta\varepsilon \qquad (1)$$

where
$$\beta = \frac{\partial \log W_R}{\partial U} \qquad (2)$$

We can now turn our attention to system $S$ which is the real focus of our interest. Suppose $S$ is in state $|j\rangle$. In order that $|ji_R\rangle$ belongs to $B'$, the basis set used for $\Omega_B$, $\eta_i$ must lie between $U-\varepsilon_j$ and $U-\varepsilon_j+\Delta$. There are $W_R(U-\varepsilon_j)\Delta$ states $|i_R\rangle$ that satisfy this condition provided $\Delta$ is small and therefore there are $W_R(U-\varepsilon_j)\Delta$ possible members of $B'$ such that $S$ is in state $|j\rangle$. Since the probability that $T$ is found to be in any one of these states is $M^{-1}$, the probability, $p_j$, that a measurement finds $S$ in state $|j\rangle$ is $W_R(U-\varepsilon_j)\Delta M^{-1}$. From equation (1)

$$p_j = M^{-1}\exp\{\log W_R(U) - \beta\varepsilon_j\}\Delta = Q^{-1}(\beta)\exp-\beta\varepsilon_j \qquad (3)$$

where $Q^{-1}$ equals $M^{-1}W_R(U)\Delta$ and is independent of $\varepsilon_j$. Since $\sum_j p_j$ must be equal to unity:

$$Q(\beta) = \sum_j \exp-\beta\varepsilon_j$$

The foregoing thus provides the derivation of equations (2.16) and (2.17) in the main text.

Having derived the canonical distribution, let us discuss the assumptions used to obtain this result. Suppose the interaction energy, $\delta$ say, when systems $S$ and $R$ are brought together is not very small. This means that the energy of state $|\psi\rangle$ is effectively changed from $\varepsilon$ to $\varepsilon + \delta$ and $p_\psi$ is changed from $Q^{-1}\exp - \beta\varepsilon$ to $Q^{-1}\exp - \beta(\varepsilon + \delta)$. $p_\psi$ is changed by a factor of order $\exp - \beta\delta$, so, for the canonical distribution to be valid, $|\beta\delta|$ must be much smaller than one.

Furthermore, if the argument presented above is to be valid, $\beta\Delta$ must also be much smaller than one. Since $T$ is a very large system this is unlikely to be true and fluctuations of its energy will be of order $N^{1/2}$ where $N$ is the number of particles in $R$. Let us assume that the probability that a measurement finds $T$ to be in an energy state between $U$ and $U + dU$ is $P(U)\,dU$. Similar arguments to those above can be applied to generalise equation (3) to become

$$p_\psi = \int_0^\infty P(U)Q^{-1}(\beta)\exp - \beta(U)\varepsilon \, dU \qquad (4)$$

Convolution arguments mentioned previously suggest that $\beta(U)$, defined as $d/dU \log W_R(U)$, in accordance with equation (2), is a function of just one variable, $(U/N)$, when $N$ is large. This means that, if variations in $U$ are of order $N^{1/2}$, variations in $\beta$ are of order $N^{-1/2}$. It can therefore be assumed that $\beta(U)$ in equation (4) can be replaced by just $\beta$ and this equation becomes

$$p_\psi = Q^{-1}(\beta)\exp - \beta\varepsilon \int_0^\infty P(U)\,dU = Q^{-1}(\beta)\exp - \beta\varepsilon$$

which is the same as equation (3). Thus the canonical distribution is proved even when realistic fluctuations in energy are allowed.

It is possible to generalise the result by allowing the very large system $R$ to be in contact with two disconnected smaller systems $S_1$ and $S_2$. Let $|\psi_1\rangle$ be an energy eigenstate of $S_1$ with eigenvalue $\varepsilon_1$ and $|\psi_2\rangle$ be an energy eigenstate of $S_2$ with eigenvalue $\varepsilon_2$. Treating $S$ as being $S_1 + S_2$, the probability that $S$ is in state $|\psi_1\psi_2\rangle$ is $Q^{-1}\exp - \beta(\varepsilon_1 + \varepsilon_2)$. The probability that $S_1$ is in state $|\psi_1\rangle$ is therefore

$$Q^{-1}\exp - \beta\varepsilon_1 \sum_{|\psi_2\rangle} \exp - \beta\varepsilon_2 = Q_1^{-1}\exp - \beta\varepsilon_1$$

for some $Q_1$, say. Similarly, the probability that $S_2$ is in state $|\psi_2\rangle$ is $Q_2^{-1}\exp - \beta\varepsilon_2$. $\beta$ is the same for the canonical distributions of both systems $S_1$ and $S_2$. Since the only thing two systems in thermal equilibrium with the same large body have in common is the temperature, $\beta$ must be a function of temperature and no other variable. As shown in the main text, $\beta = 1/kT$.

The argument can also be generalised to allow $S$ to exchange particles with $R$. The probability that $S$ is in a state with energy $\varepsilon$ and contains $n$ particles is found to be $Z^{-1}(\beta\mu)\exp-\beta(\varepsilon-\mu n)$ where $Z$ is independent of $\varepsilon$ and $n$. $\mu$ is the *chemical potential* of $S$ and is such that any other system which can freely exchange particles with $R$ has the same value of its chemical potential. $Z$ is the *grand partition function* and can be used, for example, to obtain the Bose–Einstein and Fermi–Dirac distributions. These are obtained in the main text as equation (4.25). The use of the grand partition function has the advantage of avoiding the assumption that the system is always in its 'most probable distribution' and it also avoids the adoption of the Stirling approximation.

# The statistical mechanical analogues of work and heat

The expectation value of the internal energy of some system, $S$, with Hamiltonian, $H$, is given by

$$U = \sum_j p_j \varepsilon_j \qquad (1)$$

Here the summation is over all eigenstates $|\psi_j\rangle$ of $H$, the $\varepsilon_j$ being the corresponding eigenvalues and the $p_j$ the corresponding occupation probabilities. Thus if the system is in state $|\psi\rangle$ where

$$|\psi\rangle = \sum_j \psi_j |j\rangle$$

then

$$p_j = |\psi_j|^2$$

When the macroscopic parameters of $S$, like volume or magnetic field, are changed, the Hamiltonian is also changed. When $H$ changes, the energy eigenvectors and eigenvalues also change. From equation (1)

$$dU = \sum_j p_j \, d\varepsilon_j + \sum_j \varepsilon_j \, dp_j \qquad (2)$$

It is also true that $dU = dW + dQ$ where $dW$ is the work done to the system by an external agent and $dQ$ is the heat supplied to the system from an external source. This, combined with equation (2), leads to

$$dW + dQ = \sum_j p_j \, d\varepsilon_j + \sum_j \varepsilon_j \, dp_j \qquad (3)$$

The purpose of this appendix is to show that in any reversible process

$$dW = \sum_j p_j \, d\varepsilon_j \qquad (4)$$

and

$$dQ = \sum_j \varepsilon_j \, dp_j \qquad (5)$$

thus identifying the first term of the left-hand side of equation (3) with the first term of the right-hand side and similarly for the second term.

When a body is placed in thermal contact with $S$ and removed at some later time, $H$ is unchanged since $H$ depends only on the externally imposed variables of the system, such as volume, etc. The $d\varepsilon_j$ of equation (3) are all zero. It is also obvious that no work has been done on $S$ but heat may have passed into or out of $S$. Thus in the case where $dW=0$ equation (4) is satisfied. Combining this result with equation (3), equation (5) also obtains. What has to be shown now is that the same equations hold in processes in which reversible work is done.

A process in which $S$ is in thermal isolation but its external parameters are changed infinitely slowly is called *adiabatic* and is reversible. Such a process involves a change in the Hamiltonian and the performance of work, but $dQ$, the exchange of heat, is zero. In such a process the eigenstates and eigenvalues of $H$ change gradually. In order to prove equations (4) and (5) when $dQ=0$ the adiabatic theorem is required. This states that, in an adiabatic process, $p_j$ for a non-degenerate state remains constant. Proofs of this theorem are given by Schiff (1968, p. 289) and by Messiah (1964), but unfortunately the proofs are limited to the special case where the energy eigenstates are not degenerate. As has been mentioned already, the energy eigenstates of a macroscopic system are immensely degenerate. Accordingly, we shall first use the theorem as if degeneracy does not matter and then inquire whether the allowance for degeneracy is likely to affect our conclusions about equations (4) and (5).

Consider a step in an adiabatic process in which the Hamiltonian changes from $H$ to $H+dH$, the internal energy changes from $U$ to $U+dU$ and the $j$th energy eigenvalue changes from $\varepsilon_j$ to $\varepsilon_j+d\varepsilon_j$ for each $j$. If, initially, the $p_j$ were governed by a canonical distribution, at the end of this step this will no longer be exactly true. For if for some $i,j$, $\varepsilon_i \approx \varepsilon_j$, then $p_i \approx p_j$ initially. During this step $p_i$ and $p_j$ are unchanged due to the adiabatic theorem. If $d\varepsilon_i$ is positive and $d\varepsilon_j$ is negative the new distribution will not be canonical since, according to a canonical distribution, $p_i$ should have become less than $p_j$. If we have a sequence of steps like this, of type $S_{A1}$, say, the system will end up in a state far from equilibrium, without a well-defined temperature. In order to avoid this, it must be assumed that any small step of type $S_{A1}$ is followed by a step of type $S_{A2}$, say, in which no work is done or heat is exchanged but the system is allowed to relax towards equilibrium. $S_{A1}$ combined with $S_{A2}$ will be denoted by $S_A$. In step $S_{A1}$, $dQ=0$ and each $dp_j=0$. In step $S_{A2}$, $dQ=dW=0$ and each $d\varepsilon_j=0$, so that by equation (3) $\sum_j \varepsilon_j\,dp_j=0$. Combining the results for steps $S_{A1}$ and $S_{A2}$ we conclude that, for step $S_A$, $dQ=0$ and $\sum_j \varepsilon_j\,dp_j=0$. Thus equation (5) holds and, using equation (3), it follows that equation (4) also holds.

A general infinitesimal process may be considered as a step of type $S_A$, in which $dQ = 0$, followed by a step of type $S_W$ in which $dW = 0$. Since equations (4) and (5) have been shown to hold for $S_A$ and $S_W$, they also hold for a general infinitesimal process in which both heat and work may be exchanged.

Finally, a few words about degeneracy. The adiabatic theorem is not immediately valid when there is degeneracy, or when the eigenvalues cross as $H$ is changed. However, the adiabatic theorem can easily be generalised to show that if $\alpha_\mu$ is the eigenspace with eigenvalue $\varepsilon_\mu$, $\sum_j p_{\mu j}$, where the summation is over any orthonormal set $\{|\alpha_{\mu j}\rangle\}$ spanning $\alpha_\mu$, is constant in an adiabatic process. In other words, the total occupation probability of this space is constant even though the individual $p_{\mu j}$ need not be constant. During an adiabatic process

$$
dW = dU = d\left\{ \sum_\mu \varepsilon_\mu \sum_{j=1}^{d_\mu} p_{\mu j} \right\}
$$

$$
= \sum_\mu \sum_{j=1}^{d_\mu} p_{\mu j}\, d\varepsilon_\mu + \sum_\mu \varepsilon_\mu\, d\left\{ \sum_{j=1}^{d_\mu} p_{\mu j} \right\}
$$

where $d_\mu$ is the degeneracy of the eigenspace $\alpha_\mu$. The second term of the right-hand side vanishes so that $dW = \sum_j p_j\, d\varepsilon_j$ as before. This means that degeneracy or change of degeneracy does not alter the previous result.

# *Correlation, relaxation and spin echo*

Before describing Hahn's experiment, something should be said about certain time scales which are relevant to the discussion of non-equilibrium situations. The matter can be illustrated by the example of gases, although the corresponding 'times' may be very different in liquids and solids.

A very short time in gases is the duration of collisions – the time, $c.$ $10^{-13}$ s, during which one molecule is within the force field of another. Molecules entering a collision are assumed in simple kinetic theory to have random velocities, but subsequent to collision the velocities are necessarily correlated in accordance with the requirements of momentum conservation. A longer time scale ($c.$ $10^{-10}$ s at normal pressure) is the duration of a mean free path, and it is shown in kinetic theory that the time for the traversal of a few mean free paths is sufficient for the approximate establishment of local Maxwellian equilibrium. The translational relaxation time, $t_M$, is the time for a deviation from the Maxwellian distribution of velocities to diminish to $1/e$ of its original value. And, of course, the correlations between the velocities of molecules are changed at each collision.

A much longer period, the hydrodynamic relaxation time, $t_H$, is required for equilibrium to be approximately established over the whole volume of the containing vessel and an even longer period, up to an hour or more, may be needed for temperature equalisation to be attained between gas and the walls of the container. The time scale, $t_P$, for a Poincaré recurrence is, of course, immensely greater still, of the order $10^{10^{25}}$ years for a gram mole.

The significant point we wish to make is that there are durations when a gas may not be in true equilibrium when considered from a fine grained point of view although it might appear to be so when considered in a coarser manner.

Turning now to Hahn's experiment (1950, 1952, 1953), a somewhat simplified account is as follows. A liquid material such as water or glycerol whose molecules contain hydrogen atoms is brought into a strong magnetic field in the $OX$ direction for a time sufficient to align the proton spins in that direction. At $t = 0$, a weaker radio frequency field in the $OZ$ direction is applied for a fraction of a millisecond. Its effect is to apply a torque perpendicular to the spin axes. As the result, these axes are displaced and start to precess. Due to small inhomogeneities in the field the precession velocities are not quite equal among the spins and they become more and more out of phase with each other. The macroscopic magnetic moment of the material as a whole, therefore, starts to diminish and it might appear, from a coarse grained viewpoint, that the precessions are becoming 'random'. However, from the fine grained viewpoint this is not the case, since the precession of each spin remains 'lawlike' related to the torque which acted on it. Relaxation does not occur to a significant degree and therefore true equilibrium is not attained.

At $t = t_1$, a few milliseconds later, a second radio frequency pulse is applied and its effect is to rotate the spin directions through 180° from their original directions. Provided the spins remain in the $XY$ plane (due to the smallness of the spin–spin interactions) this is equivalent to a reversal of the precessional velocities. As a consequence, the precession velocities now start to become more and more into phase with each other, the reverse of the earlier behaviour. At $t = 2t_1$ the spins have very nearly returned to the orientations they had at $t = 0$, and the overall magnetic moment has risen once again to a value close to its original value. This is the 'echo', and it can be repeated many times although with a gradually diminishing magnitude due to transfer of energy from the spins to the bulk liquid.

The echoes are thus obtained during a period analogous to the time during which, in the case of a gas, correlations between the velocities of the gas molecules still persist. The essential difference from the Loschmidt Paradox, discussed in § 3.1, is that for every member $A_i$ of the set $\{A_i\}$ of the initial precessional velocities there is a corresponding member $\bar{B}_i$ of the set $\{\bar{B}_i\}$ of inverted precessional velocities. Furthermore, the second pulse provides a mechanism for actually achieving the inversion of each $A_i$ into the corresponding $\bar{B}_i$. Therefore, there is no imbalance in the number of forward and reverse motions as there was in the case of a gas expanding into a larger volume.

Hahn illustrated the mechanism of the echo with a very beautiful analogy; at $t = 0$ a number of runners set out from a starting line on a race track and, due to their different speeds, they become distributed at positions along the track which may appear to be random. However, this is not really so if their speeds remain constant, for, in that case, their relative positions at any moment are law-bound related to each other. Therefore, if at $t = t_1$ the referee fires a second shot and, by previous arrangement, the runners quickly turn about and run in the opposite direction they will all arrive back at the starting line at the same moment $2t_1$. Here too, for every $A_i$ there is a corresponding $\bar{B}_i$ – the same person running in the opposite direction.

Similar effects can be achieved in laboratory experiments on viscous liquids. If a liquid having in it a band of colour is pushed through an assembly of tightly packed beads, the colour may appear to have become well mixed into the liquid at the outlet. However, on gently reversing the flow, the colour band can be approximately reconstituted at the former inlet point.

The question must now be raised whether the spin echo effect or the reconstitution of the colour band are phenomena which are contrary to the Second Law. That this is almost certainly not the case may be seen by comparing these phenomena with other quasi-periodic effects such as are displayed by a damped oscillator. If a pendulum is set swinging in the air from an initial position $x$, the entropy of the pendulum together with its environment steadily increases whether the pendulum swings from left to right or whether it swings back again to the left, recovering very nearly its original position $x$. There is no contravention of the Second Law and the continuous entropy increase is due to the viscous action of the air. In the case of spin echo there is a similar continued entropy creation because of the slow destruction of correlations, i.e. of the lawlike relations existing between the precessional velocities. This destruction is due to (a) the relaxation of spin energy into other forms of energy within the material, and (b) the perturbations arising from the external world. Similar considerations apply to the partial reconstitution of a colour band in a liquid.

It would appear, therefore, that although some small entropy reduction might be attributed to the partial reversal which is to be seen at the *macro*-level, this is much more than compensated for by a major effect which is the continuous entropy production occurring at the *micro*-level due to the progressive loss of correlations. Although the former process is so striking to the eye (i.e. at the coarse grained level) it is the invisible micro-process which dominates and ensures non-contravention of the Second Law.

Similar conclusions were reached by Rhim, Pines and Waugh (1971) and by Mayer and Mayer (1977). The latter pointed out that the spin echo system is never at true equilibrium during the occurrence of the echoes since the overall magnetic moment is in process of change. For this reason the thermodynamic entropy is strictly inapplicable to the system. From the standpoint of statistical mechanical entropy, account has to be taken of the degree to which correlations have decayed at any stage of the process. This requires the use of a fine grained method of approach. To use only a coarse grained method would be to obscure the difference between quasi-equilibrium, which can occur when certain relaxation processes are slow relative to observation, and true equilibrium which prevails only when relaxation is complete and when the external perturbations have had their full effect.

# The generalised H-theorem

Referring to equations (3.3) and (3.4), $\ln \bar{\rho}$, by its nature, is constant throughout a star. Furthermore, the integration of $\rho(P, t)$ over a star is the normalised value of $\bar{\rho}(\tau, P, t)$ as given by (3.3). It follows that (3.4) can be rewritten as follows at the moment $t_1$:

$$\bar{H}_1 = \int \rho_1 \ln \bar{\rho}_1 \, d\Gamma$$

Now at $t_0$ we have, from (3.5), together with (3.1),

$$\bar{H}_0 = \int \rho_0 \ln \rho_0 \, d\Gamma$$

$$= \int \rho_1 \ln \rho_1 \, d\Gamma$$

Thus

$$\bar{H}_0 - \bar{H}_1 = \int (\rho_1 \ln \rho_1 - \rho_1 \ln \bar{\rho}_1) \, d\Gamma \tag{1}$$

Since $\int \rho_1 \, d\Gamma = 1$ and $\int \bar{\rho}_1 \, d\Gamma = 1$ these two integrals may be subtracted from, and added to, equation (1) respectively to give

$$\bar{H}_0 - \bar{H}_1 = \int (\rho_1 \ln \rho_1 - \rho_1 \ln \bar{\rho}_1 - \rho_1 + \bar{\rho}_1) \, d\Gamma$$

In accordance with the well-known inequality

$$x \ln x - x \ln y - x + y \geqslant 0 \tag{2}$$

whenever $x \geqslant 0$ and $y > 0$, with strict inequality unless $x = y$, it follows that

$$\bar{H}_0 \geqslant \bar{H}_1$$

Similarly for an instant $t_2$

$$\bar{H}_0 \geqslant \bar{H}_2$$

# The free energy and entropy of mixing

## (a) Chemical potentials

The chemical potential of the $i$th species in a mixture can be defined by the partial differential:

$$\mu_i = (\partial U/\partial n_i)_{S,V,n_j}$$

As is said in Note 37, Guggenheim expressed doubts about whether this is a fully satisfactory definition. The variation $dn_i$ in the number of moles at constant entropy is an abstract idea, he said, not corresponding to any simple physical process. He went on to suggest (*loc. cit.*, p. 41) that $\mu_i$ becomes well defined by equilibrating the mixture with the pure substance $i$ through a membrane permeable only to that substance. (See also McGlashan, 1979, p. 141.)

Yet this creates the problem about whether the notion of a membrane permeable to only one substance is itself satisfactory. We have discussed this point in §4.3 in connection with Planck's method of obtaining the entropy of mixing. Guggenheim was surely well aware of this issue and it was perhaps the origin of his remark (*loc. cit.*, p. 58) that the entropy of mixing of very similar substances has to be based on statistical mechanics and not on classical thermodynamics.

It is to be noted, however, that $\mu_i$ can be equally well defined in terms of the Helmholtz free energy, $A$, or the Gibbs' free energy, $G$. Thus:

$$\mu_i = (\partial A/\partial n_i)_{T,V,n_j} = (\partial G/\partial n_i)_{T,p,n_j}$$

With these definitions $\mu_i$ is expressed in terms of the increased capacity of the system to do mechanical work per unit amount of the substance $i$ added to the system. This is much less 'an abstract idea'. (Of course, in the case of the definition in terms of $G$, the work of displacing the environment has to be excluded.)

## (b) Perfect gas mixtures and ideal solutions

If the notion of the chemical potential thus becomes fully acceptable, without the need for the invocation of semi-permeable membranes, it provides the simplest means of defining perfect gas mixtures and ideal solutions, and thereby of obtaining the free energy and entropy of mixing.

A perfect gas mixture is such that for each of its components, $i$, the change of the chemical potential at constant temperature, $T$, between two partial pressures $p_{i1}$ and $p_{i2}$ is given by

$$\mu_{i1} - \mu_{i2} = RT \ln (p_{i1}/p_{i2})$$

where $p_i \equiv x_i p$. $p$ is the total pressure of the gas mixture and $x_i$ is the mole fraction of the $i$th component. Alternatively, if $p_{i2}$ is taken to be unit pressure:

$$\mu_i = \mu_i^0(T) + RT \ln p + RT \ln x_i \tag{1}$$

where $\mu_i^0$ is a function of temperature only and is the Gibbs' free energy per mole of the pure gas $i$ at unit pressure. From this equation one can readily deduce all the properties normally associated with perfect gas mixtures such as the equation of state, $pV = RT \sum n_i$ and zero enthalpy of mixing.

The free energy of mixing is also readily obtained. The total Gibbs' free energy of the mixture is

$$G_m = \sum n_i \mu_i = \sum n_i \mu_i^0 + RT \sum n_i \ln p + RT \sum n_i \ln x_i$$

The total Gibbs' free energy of the separate gases before mixing, each at the pressure $p$ and the temperature $T$, is

$$G = \sum n_i \mu_i^0 + RT \sum n_i \ln p$$

The free energy of mixing at constant pressure and temperature is therefore

$$\Delta G_{\text{mix.}} = RT \sum n_i \ln x_i$$

It follows from the equation of state that this is also the free energy of mixing at constant total volume.

The partial molal enthalpy, $H_i$, of a component of the mixture is given by a standard thermodynamic formula:

$$\left( \frac{\partial \mu_i/T}{\partial T} \right)_{p, n_i, n_j} = -H_i/T^2$$

By applying (1) it follows that $H_i$ is independent of pressure and composition and is thus equal to the enthalpy per mole of the pure component at the given temperature. This means that the enthalpy of mixing, $\Delta H_{\text{mix.}}$, is zero, as has been said. One thus obtains for the entropy of mixing, at constant temperature and total volume,

$$\Delta S_{\text{mix.}} = -R \sum n_i \ln x_i \tag{2}$$

as in equation (4.10) of the main text.

A liquid or solid solution which is ideal over its whole range of composition can be similarly defined as one in which the chemical potential of each component is given by

$$\mu_i = \mu_i^*(T, p) + RT \ln x_i \tag{3}$$

where $\mu_i^*$ is the Gibbs' free energy per mole of the pure substance $i$ at the temperature, $T$, and the pressure, $p$, of the solution in question. Of course, (3) differs from (1) in that the pressure dependence of $\mu_i$ is not made explicit and this is necessary because the equation of state, $pV = RT \sum n_i$, does not hold for the liquid or solid states. Nevertheless, all of the accepted properties of ideal solutions, such as Raoult's law and zero enthalpy and volume change on mixing (at constant $T$ and $p$) are deducible from (3). So also is the entropy of mixing which is given by the same formula, (2), as is quoted above.

No such simple relation as (2) is available for imperfect gas mixtures or for non-ideal solutions. To be sure, the mole fractions $x_i$ in (1) and (3) may be multiplied by activity coefficients but these are essentially correcting factors and are functions of composition, as well as of temperature and pressure.

# The convergence of classical and quantal statistics

We are concerned here with the idea, raised in §4.9, that the Bose–Einstein and Fermi–Dirac statistics gradually converge to classical statistics as we pass from atomic particles to objects of a macroscopic size. It was pointed out that massive entities have an immense number of internal degrees of freedom, and that they normally also have a number of observably distinct macroscopic features which allow of distinguishability however super-ficially alike the entities appear. We shall therefore refer to their *distinguishing marks*. Scratches on their external surfaces would be but one example.

Consider an assembly of $N$ such entities and take first the case of 'uncorrected' Boltzmann counting. Here the entities are thought of as occurring in groups such that there are $y_1$ in the first group, $y_2$ in the second group, and so on. In the earliest instance of his combinatorial method, as described in §4.2, Boltzmann took the groups as differing by virtue of their energy ranges. But here the $N$ entities are taken as all having the same energy; the partitioning in terms of groups will be in regard to observably different ranges in the number of distinguishing marks on the bodies.

A partitioning into a total of $Y$ groups can be made in

$$N!/y_1! \, y_2! \dots y_Y! \quad \left( \sum_{i=1}^{Y} y_i = N \right) \tag{1}$$

different ways for a given value of the $Y$-tuple $\{y_1, y_2 \dots y_Y\}$ which will be denoted $\underline{y}$. This expression will be recognised as being similar to Boltzmann's equation (4.1) of the main text.

Now for all possible values of the $Y$-tuple the total number of 'different ways' is clearly

$$Y^N \tag{2}$$

since the first entity can be put into any one of the $Y$ groups, the second entity can also be put into any one of the $Y$ groups, and so on. This, of course, assumes the distinguishability of the entities as in classical counting.

Assuming equal *a priori* probabilities, as is appropriate if all the entities have the same energy, we obtain the following expression for the *classical* probability of the occurrence of some particular $Y$-tuple by dividing (1) by (2):

$$P_{cl.}(\underline{y}) = N!/Y^N \prod_{i=1}^{Y} y_i! \tag{3}$$

It may be noted that exactly the same result would have been obtained if (1) and (2) had both been divided by $N!$ corresponding to 'corrected' Boltzmann counting. This was the *ad hoc* procedure, described in §4.2, designed to give an additive entropy.

We turn now to the quantal statistics. Consider any one of the $N$ entities and let its wave function be written in Schrödinger notation as $\phi_\xi(\alpha)\chi_\eta(\beta)$. Here $\chi$ is a function of those macroscopic observables, denoted collectively as $\beta$, which allow of distinguishability between the $N$ entities. $\phi$, on the other hand, is a function of all other degrees of freedom, denoted collectively as $\alpha$, such as are normally considered in quantum mechanics and represent the atomic and molecular structure. The quantum numbers $\xi$ and $\eta$ specify the particular wave functions $\phi$ and $\chi$ respectively. Let $\xi$ have $X$ possible values and for the moment $X$ will be taken as finite. Similarly, let $\eta$ have $Y$ possible values.

We again consider a partitioning of the macro-observables into $Y$ groups such that there are $y_i$ entities in the $i$th group. The $Y$-tuple $\{y_1, \ldots, y_Y\}$ is again denoted $\underline{y}$. The entities are considered as being macroscopically distinguishable if they belong to different groups but not if they belong to the same group.

Now in Bose–Einstein counting the entities are distributed over the $X$ eigenstates with no restrictions on the number of entities per eigenstate. Consider the group having $y_i$ members. The number of occupied states for this group, according to the principles of the Bose–Einstein counting as described in the main text, is:

$$\frac{(y_i + X - 1)!}{y_i!\,(X-1)!} \tag{4}$$

Thus for all $Y$ groups, the number of occupied states is:

$$\prod_{i=1}^{Y} \frac{(y_i + X - 1)!}{y_i!\,(X-1)!} \tag{5}$$

Now the total possible number of symmetric states is obtained from the last expression by replacing $y_i$ by $N$ and by replacing $X$ by the product $XY$. It is therefore

$$\frac{(N + XY - 1)!}{N!\,(XY - 1)!} \tag{6}$$

The probability of the occurrence of some particular $Y$-tuple can now be obtained by dividing (5) by (6) and is:

$$P_{BE}(\underline{y}) = \frac{N!\,(XY-1)!}{(N+XY-1)!} \prod_{i=1}^{Y} \frac{(y_i+X-1)!}{y_i!\,(X-1)!} \tag{7}$$

Similarly in Fermi–Dirac statistics except that there can now be only either 0 or 1 entities per eigenstate. Following through as above, and using the results for Fermi–Dirac counting as given in the main text, we obtain for the probability of the occurrence of some particular $Y$-tuple:

$$P_{FD}(\underline{y}) = \frac{N!\,(XY-N)!}{(XY)!} \prod_{i=1}^{Y} \frac{X!}{y_i!\,(X-y_i)!} \tag{8}$$

Now the number $X$ of possible quantum states relating to the atomic and molecular structure of a macroscopic entity at fixed energy is enormously large. Consider then the limiting values of (7) and (8) as $X \to \infty$. Noticing that

$$\prod_{i=1}^{Y} X^{y_i} = X^N \quad \text{since} \quad \sum_{i=1}^{Y} y_i = N$$

it can be readily shown that the limits of (7) and (8) are

$$P_{BE}(\underline{y}) = P_{FD}(\underline{y}) = N!/Y^N \prod_{i=1}^{Y} y_i! \tag{9}$$

This is exactly the same result as was obtained in equation (3) for the classical counting.

It follows that a set of independent macroscopic objects obey the classical statistics. In particular, the tossing of a pair of coins may be expected to give on the average a pair of heads on 25 per cent of the occasions, as was asserted in the main text, and not the very different value which would be predicted on the basis that the coins behaved either as bosons or as fermions. The vast number of internal degrees of freedom of massive objects provides the reason why the classical result is obtained.

In short, there is a continuous transition from the statistics of indistinguishability to the statistics of distinguishability among entities 'of the same kind' as they go from the atomic to the macroscopic in size.

# *NOTES*

1 We refer here specifically to the measurements of macroscopic properties since it is with these that thermodynamics and statistical mechanics are concerned. Quantum measurements may, of course, be in a different category and, as is well known, a number of theoreticians hold that 'the observer' is an integral part of the overall measurement system.

2 For the derivation and conditions of validity of this equation see, for example, K. G. Denbigh (1981 *a*, pp. 45, 81).

3 In the so-called 'irreversible thermodynamics' it is assumed that equation (1.3) continues to hold *locally* under not too drastic conditions of non-equilibrium; namely when the gradients of temperature, etc., are small and when the temporal variation of thermodynamic quantities is slow compared with the slowest relaxation process in the system in question.

4 Notice that the concept of 'equilibrium' is itself an idealisation. The example of the unstable hydrogen–oxygen mixture is very familiar, and so, too, is the fact that the various chemical elements are unstable relative to iron during astronomic periods of time. The assumption of equilibrium has to be made, even when it is not strictly valid, since otherwise thermodynamics would have little practical application. And when this assumption is made, the notion of equilibrium still remains a little indefinite due to the reality of fluctuations within the supposed equilibrium state. As Landsberg (1978) has remarked, we must think, not of absolute constancy, but rather of the absence of long-term trends within the system in question. Another important point (Hobson, 1971) is that the notion of equilibrium depends on the available data. Hobson instances the case of two isotopes diffusing into each other. To an observer who is ignorant of the existence of the isotopes, the system might appear to be in equilibrium, whereas to an observer who has this knowledge the system is in process of change. But here again it is not a matter, in our view, of subjectivity but only of the adequacy of the observer's 'model'.

5 It may be noted that the entropy of isotope mixing is measurable in principle so long as reversible separation processes are available.

6 Perhaps it may be objected that what is really happening in the examples quoted is that we are using the information that 'we don't have information'! If so, this serves only to underline the ambiguous character of the whole notion of information, when used in a scientific context. The question whether certain kinds of information may reasonably be counted as negative entropy will be considered further in Chapter 5.

7 These issues are discussed in greater detail in a book by one of the present authors (K. G. Denbigh, 1981 *b*, pp. 35, 101, 124).

8 From a thermodynamic viewpoint, a system's energy cannot be measured absolutely but is always relative to a suitably chosen 'standard state'. In statistical mechanics the zero of energy is chosen so that the energy levels of a system are all non-negative.

9 For discussion of this point see Tolman (1938, p. 452); Ter Haar (1955); van Hove (1955); J. E. Mayer (1961); van Kampen (1962).

10 It should be added that in the situation which is envisaged, where there are random perturbations, the system in question is no longer describable by an exact Hamiltonian, $\hat{H}$, as is discussed in Appendix 2.1. It follows that the use of the term 'eigenstate' is also no longer exact but is an approximation.

11 See also Grad (1967), Penrose (1970, p. 43) and Jaynes (1979).

12 See also T. L. Hill (1956) and Malament & Zabell (1980).

13 The relationship between $S_{BP}$ and $S_G$ during the approach to equilibrium is discussed in an illuminating way by Klein (1956).

14 The conclusion that the $p_j$ vary exponentially with the $\varepsilon_j$ may seem inconsistent with the basic postulate to the effect that the $p_j$ are constant within the range $E$ to $E + \delta E$. However, if this range is very small, the apparent inconsistency becomes insignificant.

15 In thermodynamics, the term 'adiabatic' refers to processes in which the system in question undergoes no heat transfer. The so-called 'adiabatic theorem' is about a more restricted kind of process, one which is adiabatic *and* where the changes in the external variables are made extremely slowly. Thus it is reversible and isentropic and should really be called the 'isentropic theorem'.

16 Notice that when a body is at a uniform temperature $T$ and takes in heat $dq$, and undergoes no other change such as the performance of work, its entropy change is $dq/T$ whether or not the heat transfer occurs under reversible conditions. This is because $dq$ is then equal to $dU$.

17 It should be mentioned that Hobson (1971, p. 144) has put forward a *definition* of disorder such that it becomes tautologically true, within the context of the information theory approach to statistical mechanics, that disorder very probably increases during an irreversible process. Landsberg (1978) has given another definition.

18 In this kind of respect $S_{BP}$ offers a clearer physical understanding of entropy than does $S_G$. In the text, attention has been concentrated on the latter analogue since it is more closely related than is $S_{BP}$ to the partition function.

19 An equivalent statement is that the replacement of $t$ by $-t$ gives an equally possible solution of the equations of motion, so long as all relevant vectors (e.g. velocities, spins and magnetic fields) are reversed in direction.

20 It seems that Loschmidt, who raised his 'objection' in 1876, was unaware that Wm Thomson had already discussed velocity reversal in 1874 and had argued that it would not result in a gas returning to its original state.

21 For further discussion of the Loschmidt and Zermelo 'objections' see M. Kac (1959) and R. Balescu (1967).

22 An instance of a type of process where the number of degrees of freedom may change, is the union of two atoms to form a diatomic molecule, $2A \rightarrow B$. If the temperature is not too high, the molecule $B$ may behave as a classical dumb-

bell molecule having only five, and not six, degrees of freedom.

23 See also note 188 in P. & T. Ehrenfest (1959). Klein (1956) compares the temporal behaviour of $S_G$ and $S_{BP}$ for the urn model. The characteristics of $S_{BP}$ are also discussed in a more general sense by Penrose (1970, pp. 174 ff.). Some illuminating remarks from a thermodynamic standpoint are made by Pippard (1964).

24 S. M. Burbury (1903) raised an objection to the details of Gibbs' argument, but it seems that he did not recognise the significance of one of Gibbs' footnotes. Gibbs idealised the stirred system by regarding it in terms of continuum mechanics and not as an aggregate of molecules. Molecular diffusion was excluded. Gibbs would have made his point more clearly if he had taken the coloured, and the uncoloured, liquid as being immiscible, so that the dispersion occurs in the form of progressively smaller droplets.

25 Notice that coarse graining is not to be confused with the rule that classical statistical mechanics can be made consistent with the quantised energy states of quantum statistical mechanics by separating the allowable energy surfaces in $T$-space by volumes equal to $h^{Nf}$. The 'stars' we are here concerned with have a volume much larger than $h^{Nf}$.

26 Strictly speaking, this should be expressed as a *sum* over the stars since $\bar{\rho}$ changes in a stepwise manner, and not continuously, from one star to another.

27 Hobson (1971) has drawn attention to the fact that irreversibility occurs in macroscopic systems only when the latter approximate in their behaviour to that which would be expected in classical statistical mechanics. Those macroscopic systems, such as super-fluids and super-conductors, which display a highly quantum kind of behaviour do not manifest irreversibility.

28 By the 'normal', or fine grained, entropies we mean those given by the quantal equation (2.15) or the classical equation (3.2) when used in conjunction with the microcanonical or canonical distribution of probabilities. Landsberg (1978, p. 146) regards the use of these distributions as actually involving *a different kind* of coarse graining because they eliminate the need for data on the probabilities, $p_i$, of the individual eigenstates, or for data on the density $\rho$. However, we think it correct to regard these 'normal' entropies as being fine grained in so far as equation (2.15), for example, is a summation over *all* of the eigenstates. What we, along with most other authors, call 'coarse graining' is the dealing with the eigenstates in groups. This is effectively to do the summation over a reduced range of indices.

29 The state of the gas is much more 'unknowable' during the period of the eddies and surges than it is either in the initial state or in the final state. Therefore, if it is contended, as in information theory, that entropy is an inverse measure of information, it might appear that the entropy of the gas passes through a *maximum* before the equilibrium state is attained. Indeed, thermodynamics cannot exclude such a possibility since the eddies and surges cannot be 'frozen in' and there is no way of achieving a reversible path to such states. Nevertheless, it would be contrary to our understanding of the Second Law to suppose that the process does in fact pass through an entropy maximum. This example thus serves to cast further doubt on the reliability of interpreting entropy as 'ignorance'.

30 The assumption of the maintenance of the canonical distribution is closely

related to the assumption, used in several applications of 'non-equilibrium thermodynamics', that the Gibbs' equation:

$$dU = T\,dS - p\,dV + \sum \mu_i\,dn_i$$

continues to hold, at least locally. This appears to be a good assumption so long as local differences of temperature, for example, are small compared with the temperature itself. The matter is discussed by de Groot and Mazur (1962).

31 In general, chemical reaction does not result in *complete* conversion. Whether one starts with a mixture of $A$ and $B$, or with pure $C$, the reaction ends with an equilibrium mixture of all three substances.

32 For a contrary view see J. von Plato (1982).

33 Of course this view was held particularly strongly in *classical* statistical mechanics since this was based on a deterministic mechanics. It was not sufficiently appreciated until recently that even in classical theory there arise instabilities in the motions and these preclude the predictive form of determinism, if not the ontological form.

34 The term 'additivity', as used in this chapter, refers to the state of affairs when the entropy, or energy, of a system is the sum of the entropies, or energies, of its parts. It is important to notice that this has no necessary connection with the question whether entropy (or energy, for that matter) has an 'absolute' value, e.g. whether entropy has a natural zero at the absolute zero of temperature. Concerning the circumstances under which additivity holds, see Pippard (1964, p. 43) and Landsberg (1978, p. 77).

35 Since $Q$ is not dimensionless, Planck in 1906 proposed making it so by dividing it by $h^{3N}$ and this is the reason why $h$ makes its appearance in the constant $C$. The idea that the minimum size of 'cells' in the classical phase space is $h^{3N}$ was not fully understood until modern quantum mechanics had been developed but nevertheless was used much earlier.

36 If $\ln (2\pi N)^{1/2}$ had been included in the Stirling expression this would have resulted in a very small deviation from additivity, and in a very small positive value for $\Delta S$, in place of zero as in (4.6). This is readily understandable since a small number of relatively improbable quantum states become newly accessible when a doubled sample of gas is present in a doubled volume. This matter is discussed in §5.5.

It may be added that it is of no significance that the entropy, as given by (4.5), appears to go to $-\infty$ as $T \to 0$. The theory of the equation is only valid for perfect gases, whereas all real substances would have ceased to be perfect gases before the absolute zero of temperature is approached.

37 According to Guggenheim (1957, pp. 24 and 41), the chemical potentials of Gibbs are not well defined until one adopts the notion of a semi-permeable membrane, as used by Planck, for the purpose of equilibrating a mixture with one of its components in the pure state. But see Appendix 4.1.

38 For instance: P. W. Bridgman (1941), E. Schrödinger (1948), H. Grad (1961), D. Hestenes (1970), A. Hobson (1971), A. M. Lesk (1980).

39 The inner product was denoted $\langle \phi | \psi \rangle$ in the notation of §2.2.

40 Klein also considered the case where the energies are not the same and where the higher energy state may gradually decay into the lower energy state.

41 Siegel (1970) considers that it is necessary to quantify the minimum difference

which is necessary in order to attain distinguishability. As the minimum necessary difference, he proposes that the wave function arising from the interchange of two particles must be orthogonal to the original one. He goes on to say that non-orthogonal states are 'not entirely different from one another'.

42 See also Lyuboshits & Podgoretskii (1971), who claim that Landé's equation contains an error.

43 For further discussion on this point see Lesk (1980).

44 There has been lively controversy on the applicability of Leibniz' Principle to elementary particles between Cortes (1976), Barnette (1978), Ginsberg (1981) and Teller (1983).

45 It has been claimed that Boltzmann made an implicit use of the concept of transcendental individuality.

46 Gibbs seems to have anticipated that cardinal principle of the American school of philosophical pragmatism: in connection with any philosophical issue one should always ask 'What difference does it make?'

47 See, for example, Mott (1930), Mott & Massey (1965) and Feynman & Hibbs (1965).

48 Greenberg & Messiah (1964, 1965), Girardeau (1965), Hartle & Taylor (1969), Stolt & Taylor (1970).

49 Although Einstein predicted the boson 'condensation' phenomenon, he expressed his puzzlement that the differences between classical and quantal counting 'express indirectly a certain hypothesis on a mutual influence of the molecules which for the time being is of a quite mysterious nature'. (Quoted from Pais, 1979.)

50 A useful treatment of combinatorial problems is given by Mayer & Mayer (1940).

51 It will be seen that the numerical results quoted in the example are in accordance with the general expressions (4.19), (4.21) and (4.23). For the case $g_\kappa = 2$ and $n = n_\kappa = 2$ these expressions (remembering that $0! = 1$) yield the numbers 3, 1 and 4 respectively. The corresponding probabilities are thus $\frac{1}{3}$, 1 and $\frac{1}{4}$ respectively.

52 See also Weinreich (1959) and Sudersham & Mehra (1970).

53 Concerning 'identity' in the field of philosophy see, for example, Wiggins (1980) and Brody (1980), as well as the books by Strawson and Quinton already referred to.

54 In a partly autobiographical article (1979) Jaynes recounts how he agonised over this problem. In the context of Shannon's application he came to the conclusion that $H$ measures the 'ignorance' of the engineer who originally designed the equipment of the channel since that channel must be able to cope, during its useful life, with the set of all messages which *might* be sent through it. This leaves unanswered how other contexts might be dealt with.

55 The first inequality arises from the fact that $W_0$ refers to *all* of the molecules in the enclosure whereas $\delta W$ refers to the effect of transferring only one molecule. The second inequality has already been mentioned: $h v_f$ must exceed $k T_0$.

56 It may be added that a more detailed treatment of the Demon problem shows that consistency with the Second Law obtains only *on the average*, i.e. as an average over many repetitions.

57 It is of interest to compare an objective physical situation with an individual's betting on a horse race. Suppose a punter represents his estimate of the probability distribution of the various horses winning by means of the $H$ function. There would clearly be nothing peculiar if his value of $H$ were either to increase or to diminish, according to how his opinions about the horses change during a period of time. No question would arise about these changes of $H$ being in accordance with the Second Law since they are not related to changes of thermodynamic entropy.

58 This result is an instance of Khinchin's third condition for the uniqueness theorem to hold, but is here derived from the definition of $S_G$ rather than being regarded as an axiom.

# REFERENCES

Aston, J. G. 1942. In *A Treatise on Physical Chemistry*, ed. H. S. Taylor & S. Glasstone. Van Nostrand.

Balescu, R. 1967. *Physica*, **36**, 433.

Band, W. & Park, J. L. 1976. *Found. of Phys.*, **6**, 157, 249.

Band, W. & Park, J. L. 1977. *Ibid.*, **7**, 233, 705.

Barnette, R. L. 1978. *Phil. of Sci.*, **45**, 466.

Bekenstein, J. D. 1973. *Phys. Rev.*, **D7**, 2333.

Bergman, P. G. & Lebowitz, J. L. 1955. *Phys. Rev.*, **99**, 578.

Bergman, P. G. & Lebowitz, J. L. 1959. *Annals of Phys.*, **1**, 1.

Berry, M. V. 1978. *Am. Inst. of Phys. Conf. Proc.*, No. 46, p. 16.

Blatt, J. M. 1959. *Prog. Theor. Phys.*, **22**, 745.

Boltzmann, L. 1877. *Kais. Acad. d. Wiss. Sitz.*, **76**, 373.

Borel, E. 1928. *Le Hasard*. Alcan, Paris.

Born, M. 1949. *Natural Philosophy of Cause and Chance*, p. 72. Oxford Univ. Press.

Bridgman, P. W. 1941. *The Nature of Thermodynamics*. Harvard Univ. Press.

Bridgman, P. W. 1950. *Rev. Mod. Phys.*, **22**, 56.

Brillouin, L. 1953. *J. Appl. Phys.*, **24**, 1152.

Brillouin, L. 1954. *Ibid.*, **25**, 595.

Brillouin, L. 1962. *Science and Information Theory*. Academic Press.

Brody, B. A. 1980. *Identity and Essence*. Princeton Univ. Press.

Brush, S. G. 1976. *The Kind of Motion We Call Heat*, p. 590. North-Holland Publ. Co.

Burbury, S. M. 1903. *Phil. Mag.*, **6**, 251.

Cortes, A. 1976. *Phil. of Sci.*, **43**, 491.

Cox, R. T. 1955. *Statistical Mechanics of Irreversible Change*. Johns Hopkins Press, Baltimore.

Cyranski, J. F. 1978. *Found. of Phys.*, **8**, 493.

de Beauregard, O. Costa & Tribus, M. 1974. *Helv. Phys. Acta*, **47**, 238.

de Groot, S. R. & Mazur, P. 1962. *Non-Equilibrium Thermodynamics*. North-Holland Publ. Co.

Demers, P. 1944. *Can. J. Res.*, **22**, 27.

Demers, P. 1945. *Ibid.*, **23**, 47.

Denbigh, K. G. 1981 *a*. *The Principles of Chemical Equilibrium*, 4th edn. Cambridge Univ. Press.

Denbigh, K. G. 1981 *b*. *Three Concepts of Time*. Springer-Verlag.

d'Espagnat, B. 1976. *Conceptual Foundations of Quantum Mechanics*, 2nd edn. W. A. Benjamin, Reading, Mass.

Dirac, P. A. M. 1926. *Proc. Roy. Soc.*, **A112**, 661.

Dirac, P. A. M. 1958. *The Principles of Quantum Mechanics*, 4th edn, p. 207. Oxford Univ. Press.

Ehrenfest, P. & T. 1912. *The Conceptual Foundations of the Statistical Approach in Mechanics*, English transl. 1959. Cornell Univ. Press.

Fano, U. 1983. *Rev. Mod. Phys.*, **55**, 805.

Farquhar, I. E. 1964. *Ergodic Theory in Statistical Mechanics*. Interscience.

Fast, J. D. 1970. *Entropy*, 2nd edn. Macmillan.

Feinstein, A. 1958. *Foundations of Information Theory*. McGraw-Hill.

Feynman, R. P. & Hibbs, A. R. 1965. *Quantum Mechanics and Path Integrals*, p. 14. McGraw-Hill.

Fowler, R. H. & Guggenheim, E. A. 1939. *Statistical Thermodynamics*. Cambridge Univ. Press.

Gibbs, J. W. 1876. From *The Scientific Papers*, vol. 1, p. 166–7. Dover, New York, 1961.

Gibbs, J. W. 1902. *Elementary Principles in Statistical Mechanics*. Yale Univ. Press. (All quotations are from the Dover edn, 1960.)

Gillies, D. A. 1973. *An Objective Theory of Probability*. Methuen.

Ginsberg, A. 1981. *Phil. of Sci.*, **48**, 487.

Girardeau, M. D. 1965. *Phys. Rev.*, **139**, B 500.

Grad, H. 1961. *Comm. Pure and Appl. Maths*, **14**, 323.

Grad, H. 1967. *Delaware Seminar in The Foundations of Physics*, ed. M. Bunge. Springer-Verlag.

Greenberg, O. W. & Messiah, A. M. L. 1964. *Phys. Rev.*, **136**, B 248.

Greenberg, O. W. & Messiah, A. M. L. 1965. *Ibid.*, **138**, B 1155.

Griffiths, R. B. 1965. *J. Math. Phys.*, **6**, 1447.

Grünbaum, A. 1975. In *Entropy and Information*, ed. L. Kubát & J. Zeman. Academia, Prague.

Guggenheim, E. A. 1949. *Research*, **2**, 450.

Guggenheim, E. A. 1957. *Thermodynamics*, 3rd edn. North-Holland Publ. Co.

Hacking, I. 1965. *The Logic of Statistical Inference*. Cambridge Univ. Press.

Hahn, E. L. 1950. *Phys. Rev.*, **80**, 580.

Hahn, E. L. 1952. *Ibid.*, **88**, 1070.

Hahn, E. L. 1953. *Physics Today*, **6**, No. 11, p. 4.

Hanle, P. A. 1977. *Archive for Hist. Exact Sci.*, **17**, 165.

Hartle, J. B. & Taylor, J. R. 1969. *Phys. Rev.*, **178**, 2045.

Hawking, S. W. 1976. *Phys. Rev.*, **D13**, 191.

Hestenes, D. 1970. *Am. J. Phys.*, **38**, 840.

Hill, T. L. 1956. *Statistical Mechanics*. McGraw-Hill.

Hobson, A. 1971. *Concepts in Statistical Mechanics*. Gordon & Breach, New York.

Hoyningen-Huene, P. 1976. *Physica*, **82A**, 417; **83A**, 584.

Jancel, R. 1969. *Foundations of Classical and Quantum Statistical Mechanics*. Pergamon.

Jauch, J. M. & Báron, J. G. 1972. *Helv. Phys. Acta*, **45**, 220.

Jaynes, E. T. 1957. *Phys. Rev.*, **106**, 620; **108**, 171.

Jaynes, E. T. 1963. In *Information Theory and Statistical Mechanics*, ed. K. W. Ford. Benjamin, New York.

Jaynes, E. T. 1965. *Am. J. Phys.*, **33**, 391.

Jaynes, E. T. 1971. *Phys. Rev.*, **A4**, 747.

Jaynes, E. T. 1979. In *The Maximum Entropy Formalism*, ed. R. D. Levine & M. Tribus. M.I.T. Press.

Jaynes, E. T. 1981. Paper presented to the First ASSP Workshop on Spectral Estimation. McMaster University.

Kac, M. 1959. *Probability and Related Topics in Physical Science.* Interscience.

Kastler, A. 1983. In *Old and New Questions in Physics, Cosmology and Theoretical Biology,* ed. A. van der Merwe. Plenum Press, New York.

Khinchin, A. I. 1957. *Mathematical Foundations of Information Theory.* Dover, New York.

Klein, M. J. 1956. *Physica,* **22,** 569.

Klein, M. J. 1958. *Proc. Ned. Acad. Wet.,* **62,** 41, 51.

Klein, M. J. 1959. *Ned. Tid. voor Natuurkunde,* **25,** 73.

Kubo, R. 1965. *Statistical Mechanics.* North-Holland Publ. Co.

Kuhn, T. 1978. *Black-Body Theory and The Quantum Discontinuity, 1894–1912,* p. 45. Clarendon Press.

Kyburg, H. E. 1970. *Probability and Inductive Logic,* Ch. 3. Macmillan.

Kyburg, H. E. 1974. *Brit. J. Phil. Sci.,* **25,** 358.

Landé, A. 1965. *New Foundations of Quantum Mechanics.* Cambridge Univ. Press.

Landsberg, P. T. 1978. *Thermodynamics and Statistical Mechanics.* Oxford Univ. Press.

Lavis, D. 1977. *Brit. J. Phil. Sci.,* **28,** 255.

Layzer, D. 1976. *Astrophys. J.,* **206,** 559.

Lesk, A. M. 1980. *J. Phys. (A. Math. Gen.),* **13,** L111.

Lewis, G. N. 1930. *Science,* **LXXI,** 569.

Lindblat, G. 1973. *Comm. Math. Phys.,* **33,** 305.

Lindblat, G. 1974. *J. Stat. Phys.,* **11,** 231.

Lyuboshits, V. I. & Podgoretskii, M. I. 1971. *Soviet Phys. Dok.,* **15,** 858, 1022.

Malament, D. B. & Zabell, S. L. 1980. *Phil. of Sci.,* **47,** 339.

Mandl, F. 1974. *Statistical Physics.* Wiley.

Maxwell, J. C. 1875. *Theory of Heat,* p. 153.

Mayer, J. E. 1961. *J. Chem. Phys.,* **34,** 1207.

Mayer, J. E. & Mayer, M. G. 1940. *Statistical Mechanics.* Wiley, New York.

Mayer, J. E. & Mayer, M. G. 1977. *Ibid.,* 2nd edn. Wiley Interscience.

McGlashan, M. L. 1979. *Chemical Thermodynamics.* Academic Press.

Mehlberg, H. 1980. *Time, Causality and The Quantum Theory.* Reidel Publ. Co., Dordrecht.

Meixner, J. 1941. *Ann. d. Physik,* **39,** 333.

Mellor, D. H. 1971. *The Matter of Chance.* Cambridge Univ. Press.

Mellor, D. H. 1983. *Brit. J. Phil. Sci.,* **34,** 97.

Melsheimer, O. 1982. *Found. of Phys.,* **12,** 59.

Messiah, A. 1964. *Quantum Mechanics,* vol. 2. North-Holland Publ. Co.

Mott, N. F. 1930. *Proc. Roy. Soc.,* **A126,** 259.

Mott, N. F. & Massey, H. S. W. 1965. *The Theory of Atomic Collisions,* Ch. XI. Oxford Univ. Press.

Newton, I. 1730. *Opticks,* 4th edn, p. 373.

Ono, S. 1949. *Mem. Fac. Eng. Kyusho Univ.,* **11,** 125.

Pais, A. 1979. *Rev. Mod. Phys.,* **51,** 861.

Penrose, O. 1970. *Foundations of Statistical Mechanics.* Pergamon Press.

Penrose, O. 1979. *Rep. Prog. Phys.,* **42,** 1937.

Percival, I. C. 1961. *J. Math. Phys.,* **2,** 235.

Percival, I. C. 1962. *Ibid.,* **3,** 386.

Pippard, A. B. 1964. *The Elements of Classical Thermodynamics*. Cambridge Univ. Press.

Planck, M. 1927. *Treatise on Thermodynamics*, 3rd English edn. Longmans, Green: London.

Popper, K. R. 1956. *Brit. J. Phil. Sci.*, **10**, 25.

Popper, K. R. 1974. In *The Philosophy of Karl Popper*, vol. 1, p. 130, ed. P. A. Schilpp. Open Court.

Popper, K. R. 1982. *Quantum Theory and The Schism in Physics*. Hutchinson, London.

Popper, K. R. 1983. *Realism and The Aims of Science*. Hutchinson, London.

Post, H. 1963. *The Listener*. October 10.

Prigogine, I. 1949. *Physica*, **15**, 272.

Prigogine, I. 1980. *From Being to Becoming*, p. 12. W. H. Freeman, San Francisco.

Prigogine, I., George, C., Henin, F. & Rosenfeld, L. 1973. *Chemica Scripta*, **4**, 5.

Quinton, A. 1978. *The Nature of Things*. Routledge and Kegan Paul.

Redhead, M. L. 1983. *Phil. of Sci. Assocn. Meeting, 1982*, vol. 2, p. 57, ed. P. D. Asquith & T. Nickles. East Lancing, Mich.

Reichenbach, H. 1956. *The Direction of Time*. Univ. of California Press.

Reif, F. 1965. *Fundamentals of Statistical and Thermal Physics*. McGraw-Hill.

Rhim, W.-K., Pines, A. & Waugh, J. S. 1971. *Phys. Rev.*, **B3**, 684.

Robertson, H. S. & Huerta, M. A. 1970. *Proc. Int. Conf. on Thermodynamics*, ed. P. T. Landsberg. Butterworths, London.

Rosenfeld, L. 1953. *Science Progress*, **41**, 393.

Rowlinson, J. S. 1963. *The Perfect Gas*. Pergamon Press.

Schiff, L. I. 1968. *Quantum Mechanics*, 3rd edn. McGraw-Hill.

Schrödinger, E. 1948. *Statistical Thermodynamics*. Cambridge Univ. Press.

Shannon, C. E. & Weaver, W. 1949. *The Mathematical Theory of Communication*. Univ. of Illinois Press, Urbana.

Siegel, A. 1970. *Found. of Phys.*, **1**, 57.

Skagerstam, B. K. 1975. *J. Stat. Phys.*, **12**, 449.

Sklar, L. 1977. *J. of Phil.*, **74**, 494.

Sterne, O. 1949. *Rev. Mod. Phys.*, **21**, 534.

Stolt, R. H. & Taylor, J. R. 1970. *Nuclear Phys.*, **B19**, 1.

Strawson, P. F. 1959. *Individuals*. Methuen.

Sudersham, E. C. G. & Mehra, J. 1970. *Int. J. Theor. Phys.*, **3**, 245.

Teller, P. 1983. *Phil. of Sci.*, **50**, 309.

Ter Haar, D. 1955. *Rev. Mod. Phys.*, **27**, 289.

Thomson, W. 1874. *Proc. Roy. Soc. Edin.*, **8**, 325, re-printed in S. G. Brush, *Kinetic Theory*. Pergamon Press.

Tisza, L. 1966. *Generalized Thermodynamics*. M.I.T. Press.

Tolman, R. C. 1938. *The Principles of Statistical Mechanics*. Oxford Univ. Press.

Tribus, M. 1963. *Boelter Anniversary Volume*. McGraw-Hill.

van Hove, L. 1955. *Physica*, **21**, 517.

van Hove, L. 1957. *Rev. Mod. Phys.*, **29**, 200.

van Kampen, N. G. 1962. In *Fundamental Problems in Statistical Mechanics*, ed. E. G. D. Cohen *et al.* North-Holland Publ. Co.

von Neumann, J. 1955. *Mathematical Foundations of Quantum Mechanics*, English transl. Princeton Univ. Press.

von Plato, J. 1982. *Phil. of Sci.*, **49**, 51.

Watanabe, S. 1969. *Knowing and Guessing*. Wiley.

Weinreich, G. 1959. *Nature*, **184**, 1825.

Wiggins, D. 1980. *Sameness and Substance*. Blackwells, Oxford.

Worrall, J. 1982. *Phil. Quart.*, **32**, 201. (See also *Synthese*, 1982, **51**, p. 141 ff.)

Yourgrau, W., van der Merwe, A. & Raw, G. 1966. *Treatise on Irreversible and Statistical Thermophysics*. Macmillan.

Yvon, J. 1969. *Correlations and Entropy in Classical Statistical Mechanics*. Pergamon Press.

Zeh, H. D. 1970. *Found. of Phys.*, **1**, 69.

# INDEX